Flammable Hazardous Material

Flammable Hazardous Material

Third Edition

SCOTT W. KENLEY
City of Brisbane Fire Department

JAMES H. MEIDL

PRENTICE HALL, Englewood Cliffs, New Jersey 07632

blication Data

tt W. Kenley, James H. Meidl. -

   p.   cm.
   Rev. ed. of: Flammable hazardous material / James H. Meidl. 2nd
ed. 1978.
   Includes bibliographical references and index.
   ISBN 0-02-380136-0
   1. Hazardous substances. 2. Inflammable materials. I. Meidl,
James H. II. Meidl, James H. Flammable hazardous material.
III. Title.
T55.3.H3M4 1995
604.7—dc20   94-48509
                                                        CIP

Publisher: Susan Katz
Production Supervisor: Helen Wallace
Production Manager: Francesca Drago
Director of Production/Manufacturing: Bruce Johnson
Text Designer: Robert Freese
Cover Designer: Rosemarie Votta

 © 1995 by Prentice-Hall, Inc.
A Simon & Schuster Company
Englewood Cliffs, New Jersey 07632

The author and publisher of this book have used their best efforts in preparing this
book. These efforts include the development, research, and testing of the theories
and programs to determine their effectiveness. The author and publisher shall not
be liable in any event for incidental or consequential damages in connection with,
or arising out of, the furnishing, performance, or use of these programs.

PRINTED IN THE UNITED STATES OF AMERICA

10 9 8 7 6 5 4 3 2 1

ISBN 0-02-380136-0

Prentice-Hall International (UK) Limited, *London*
Prentice-Hall of Australia Pty. Limited, *Sydney*
Prentice-Hall Canada Inc., *Toronto*
Prentice-Hall Hispanoamericana, S.A., *Mexico*
Prentice-Hall of India Private Limited, *New Delhi*
Prentice-Hall of Japan, Inc., *Toyko*
Simon & Schuster Asia Pte. Ltd., *Singapore*
Editora Prentice-Hall do Brasil, Ltda., *Rio de Janeiro*

Contents

Preface

Those who are responsible for protecting human life and property from fire must face the consequences of progress. New industrial materials are being introduced at an ever-increasing rate to satisfy the requirements of modern civilization. Many of these materials have properties that make them dangerous. Firefighters must recognize and combat these hazards.

In the late 1960s, James Meidl, a fire science instructor, wrote two textbooks designed to present the basics of fire chemistry and the properties and hazards of chemicals faced by firefighters. Those two books were *Flammable Hazardous Materials* and *Explosive and Toxic Hazardous Materials*. *Flammable Hazardous Materials* was first published in 1970 and revised in 1978. He felt that far too many firefighters were dying because of ignorance relative to hazardous materials and chemical reactions. Much of what James wanted to impart to his brethren is still applicable today.

This revision includes updated material and introduces a new more in-depth chapter on the structure and relationships of atoms. It is not meant to be a primer on the chemistry of hazardous materials, but a relatively easy-to-read, understandable text on the potential hazards faced by emergency services personnel during their career. Many of the updated incidents still result in the loss of life by civilians and emergency personnel. However, the preventable loss of life due to lack of knowledge has been greatly reduced over the past decade. I believe that James Meidl has played a significant role in the reduction of unnecessary life loss due to this lack of knowledge.

When I entered the fire service in 1971, *Flammable Hazardous Materials* and *Explosive and Toxic Materials* were the first two textbooks I read as a rookie firefighter. When I was asked to write the third edition of *Flammable Hazardous Materials,* I was honored at being able to continue what James had started. James Meidl passed away in the mid-1980s; however, his words continue to educate and save lives. It has been my pleasure to revise his work so that his contribution to the fire service still has meaning twenty years later.

This book is dedicated to the memory of James Meidl, his widow Helene whose devotion to James and his ideals resulted in this revision, and my wife Barbara, whose support and understanding helped me through the difficulties.

S.W.K.

Flammable Hazardous Material

Physical Properties of Hazardous Materials

A material may be toxic, reactive, and/or explosive given certain circumstances. However, the first question usually asked by a firefighter is, "Does it burn?" This philosophy is the cornerstone of most fire recruit training programs. Yet, many fire department operations and emergency situations have nothing to do with fire. In fact, many organizations are renaming the fire suppression section "Emergency Services." Therefore, the hazards of any material encountered, whether it be flammability, reactivity, toxicity, or a combination of the group, are significant. To accurately assess potential hazards faced in the field, an understanding of the basic principles of chemistry must be developed.

The knowledge learned in high school or college chemistry is usually forgotten within two or three years, primarily through lack of use. Some firefighters have not been exposed to the science of chemistry. An understanding of chemical terms, concepts, and definitions is necessary to accurately interpret the information contained in hazardous materials manuals and reference materials. Therefore, a basic understanding of atomic structure, chemical reaction, compound formation, and chemical equations is required. Your ability to draw on the information presented could, at some point in your career, save your life and those of fellow workers.

All matter can be divided into three forms: solid, liquid, and gas. All matter that is flammable burns in its gaseous form. This scientific fact requires that all flammable matter be heated sufficiently to change its physical form, resulting in the release of flammable gases. The heating and release of gases are primarily responsible for the other related hazards of reactivity and toxicity. Because heat tends to create the other hazards, the chemistry of fire is the logical place to start the discussion of flammable hazardous materials.

A good introduction to questions about flammability is a discussion of the **fire tetrahedron** (Figure 1-1). Until the late sixties, it was an accepted belief by firefighters that a combination of three factors—**heat, fuel,** and **oxygen**—was necessary before combustion could occur. Remove one of the key ingredients and the fire went out. However, this theory was not supported by the fact that fires can be extinguished with dry chemical powders, carbon tetrachloride, and chlorinated hydrocarbon gases. Because these extinguishing agents do not remove the fuel, dissipate heat, or exclude oxygen, how do they extinguish fire? Investigation into the chemistry of fire has identified the existence of a fourth factor, the **chemical chain reaction**. These extinguishing agents act to interrupt the chemical chain reaction that exists during combustion.

The firefighter's role is to apply firefighting techniques to separate the fuel from the heat, reduce the heat, exclude oxygen, or break up the chemical chain reaction. To effectively accomplish this role, the firefighter must develop an understanding of how fire burns. The remainder of this chapter is devoted to the basic definitions that relate to the **heat, fuel,** and **oxygen** sides of the fire tetrahedron. Chapter 2 will present the chemical knowledge required to understand fully the chemical chain reaction aspect of combustion.

HEAT

The firefighter's work is directly related to the basic principles of heat. Dissipating unwanted heat is one way to accomplish fire extinguishment. The generation of heat causes many of our problems.

Heat Energy

Heat energy, our main concern in this book, can be produced by: (a) The accumulation of molecules—composition and polymerization; (b) The

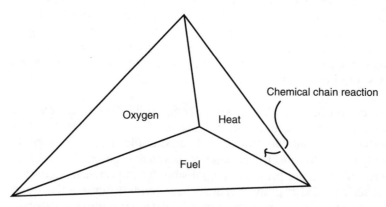

FIGURE 1–1 A fire tetrahedron.

breaking apart of molecules—decomposition; and/or (c) the heat of solution when materials are dissolved in a liquid, to name a few. Energy is defined as the ability to do work.

Mechanical heat energy is by produced by friction, two materials rubbing together. Air molecules in the compression of a gas, and application of automobile brakes are a few examples.

Electrical heat energy is produced by lightning, arcing wires, the discharge of static electricity, electrical resistance, and so on.

Nuclear heat energy is produced by fission, as in the atom bomb; fusion, as in the hydrogen bomb; or a combination of both.

Fission is the splitting apart of very heavy atoms such as Uranium-235 into lighter atoms with a resulting loss of mass or weight and the conversion of this lost substance into energy. It is this type of energy that is utilized to heat water into steam to drive steam turbines that generate electricity in nuclear power plants.

Fusion is the ramming together of very light atoms such as deuterium and tritium, alternate forms of hydrogen, producing heavier atoms, and again releasing tremendous amounts of energy in the form of heat. Nuclear energy is created on the sun by nuclear fusion.

Heat Transfer

According to the laws of physics, heat tends to flow from a hot substance or place to a cold substance or place. This fact explains the ability of a material to absorb heat from another. Heat transfers four different ways:

1. **Direct contact.** The direct contact of a heat source on another object. A pot, heated by the burner on a stove, is an example of heat transfer by direct contact. Whether the heat source is an open flame or electric element is immaterial. To qualify as heat transfer by direct contact, the heat must be provided by a heat source, not an object through which the heat is being transferred.
2. **Conduction.** The transfer of heat *through a medium*. Metals are the best conductors of heat. Heat transfer by conduction would occur if a pot, heated on a stove, burned the person who picked it up. The heat from the burner on the stove was transferred through the pot to the individual.
3. **Convection.** The transfer of heat *with a medium*. This type of heating is usually circulatory. A medium, such as air, becomes heated and moves to a cooler area, gradually heating that area. This process continues until all heat is transferred, equalizing the temperature of the entire area. The phenomenon known as flashover is a graphic example of heat transfer by convection.
4. **Radiation.** The transfer of heat *which is not dependent upon any medium*. The heat that travels from the sun to the earth through a

vacuum is an excellent illustration of this method of heat transfer. Glass, which is a poor conductor of heat, passes radiated heat quite readily.

Heat versus Temperature

What is heat energy? Heat energy has two yardsticks by which it is measured: temperature and heat. The relationship of volts versus amps is the electrical equivalent of heat versus temperature.

Molecules, either as elements or compounds, are constantly in motion. Molecules in solid material vibrate very slowly, so slowly that they remain locked in position. As energy in the form of heat is applied, the vibratory rate of the molecules increases to the point at which they can no longer maintain their locked positions. Molecules begin to slide over one another yet remain in contact, and the solid changes its physical state by melting into a liquid. As additional heat energy is absorbed, the molecules in the liquid begin to vibrate so rapidly that they separate from one another, flying off in all directions—resulting in the liquid's boiling and forming a gas.

Heat is defined as the total amount of vibration in a group of molecules and is measured relative to absorption of heat energy. The quantity of heat absorbed is measured in one of two ways:

British Thermal Unit (BTU): The amount of heat required to raise the temperature of one pound of water one degree Fahrenheit.

Calories: The amount of heat required to raise the temperature of one gram of water one degree Centigrade.

Temperature is defined as the measurement of how fast molecules are vibrating. Temperature is measured one of two ways: degrees Fahrenheit and degrees Centigrade. Conversion of one to the other is:

Degrees Fahrenheit=(9/5)°C+32

Degrees Centigrade=5/9(°F-32)

Note: Hot objects vibrate at a high rate of speed. Burns result when an individual touches a hot object and the hot object transfers its vibratory rate to the skin, causing pain and damage to the skin's delicate structure. A burn can also result from constant exposure to a relatively low temperature if the object is in contact with the heat source for an extended period of time. The key is the total absorption of heat, whether acquired over milliseconds, minutes, hours, days, or years. Once a set amount of heat is absorbed, the resultant decomposition is the same. This cumulative effect of heat absorption is known as *pyrolysis*.

Specific Heat

Specific heat is the ratio between the amount of heat necessary to raise the temperature of a material one degree Fahrenheit and the amount of heat necessary to raise the same weight of water one degree Fahrenheit.

In effect, specific heat is a comparison. Water is chosen as the standard because it is convenient. Therefore, the specific heat of water is one. It takes more heat to raise the temperature of water than it does almost any other substance. Most substances will have a specific heat of less than one, since the amount of heat required to raise the temperature of one pound of almost any other substance will be less than one BTU.

Latent Heat

Latent heat is the amount of heat a material absorbs when it changes physical form: solid to a liquid (latent heat of fusion), liquid to a gas (latent heat of vaporization), and vice versa. Water's high **latent heat of vaporization** gives it much of its firefighting effectiveness. Because we rarely shovel snow or ice onto a fire, **latent heat of fusion** is of limited importance to the firefighter. In either case water has an advantage; its latent heat properties are substantially higher than those of other substances.

Only 180 BTUs are required to raise the temperature of one pound of water 180 degrees Fahrenheit, from 32°F to 212°F. This same pound of water, as it turns to steam, will absorb an additional 970 BTUs. An enormous amount of heat energy is absorbed without any equivalent temperature rise. This energy is all used to turn sliding molecules into flying molecules. Water's high latent heat of vaporization allows for the utilization of fog stream application in the form of fine droplets to get the maximum cooling effect.

Melting and Boiling Points

All substances or groups of molecules, at some given temperature and pressure, exist in all three physical states: solid, liquid, and gas. The physical structure of elements, molecules, and compounds varies, causing them to melt or boil at different rates of vibration. Oxygen is a solid at temperatures less than -361°F, a liquid between –361°F and –297°F, and a gas above –297°F. A water molecule requires a much higher vibratory rate to melt, 32°F (0°C), and boils at 212°F (100°C). Iron requires a tremendous rate of vibration to melt: 2,795°F (1,535°C).

Note: Gases that are cooled or pressurized to their liquid and/or solid states are called **cryogenics**. The properties and handling of cryogenics will be discussed in Chapter 7.

FUELS

As stated earlier, all matter burns in a gaseous physical state. Because gases are already in that physical state, they become the most dangerous if they are flammable. Also, some liquids, such as gasoline, emit sufficient vapors at ambient temperature to produce flammable gases. Therefore, even though they appear to be a liquid, their flammability and hazard dangers are the same as those of gases. Finally, solids usually require heat from some external force significant enough to decompose the solid material into liquid and gases. The flammable gas given off in the decomposition process becomes the fuel, not the solid itself. The chemical properties of fuels will be discussed in Chapter 2. However, certain physical properties of fuels affect flammability.

Ignition Temperature

Ignition temperature is the minimum temperature required to initiate or cause self-sustained combustion of a material independent of the heat source. Ignition temperature applies to all flammable substances regardless of physical form. The key factor is the temperature of the material. It is relatively easy to heat gas molecules to their ignition temperature almost simultaneously with the application of a heat source with sufficient temperature. However, solids require a more substantial heat source. For example, light a match in the presence of the right mixture of natural gas, and there is immediate ignition; take the same match and place it at the end of a two-by-four piece of wood, and ignition is difficult if not impossible.

Ignition temperature with respect to a solid is applied based on the amount of heat absorbed as discussed in the section on heat. The significance of ignition temperature relative to ordinary combustibles is evident in fires that are burning out of control. Pre-ignition of materials ahead of a brush or wildland fire and pre-ignition of flashover are graphic examples.

Ignition temperature with respect to flammable liquids is applied as it relates to the ignition source. Flammable liquids are producing flammable gases long before the liquid itself reaches its ignition temperature. Therefore, as long as a flammable liquid is beyond its flash point, the ignition temperature of a flammable liquid will dictate the heat of the ignition source. For example , a smoldering cigarette burns at a temperature lower than the ignition temperature of gasoline and does not readily ignite gasoline fumes when a match does. A flammable gas's relationship to ignition temperature is the same as that of flammable liquids. Flammable gases are always producing sufficient quantities of vapors to be ignited by a sufficient ignition source.

Ignition Source

Any ignition source must be at least as hot as the ignition temperature of the material. A flame must be hot enough to heat a substance to its ignition temperature in the presence of air. Sparks must be intense enough and last long enough to ignite flammable vapor–air mixtures. Hot surfaces must be large enough, contain sufficient heat, and remain in contact with the material for a sufficient length of time for the material to reach its ignition temperature.

Generally, the most common ignition sources have a temperature much higher than the ignition temperature of most flammable substances. A match flame is over 2,000°F (1,093°C); electric arcs are about the same. The temperature of a cigarette ranges from 550°F to 1,350°F (288°C to 732°C).

Flash Point

Flash point is the temperature at which a flammable liquid produces enough vapors to be ignited. To the firefighter, this is the most important property of a flammable liquid. Flash point is not to be confused with ignition temperature. Flash point temperatures are not ignition temperatures; in fact, they are considerably lower than a material's respective ignition temperature. Even though a flammable liquid may be above its flash point, it still requires an ignition source with a temperature equal to or higher than its ignition temperature.

Flammable Range

Situations often occur in which flammable vapors from a low flash point liquid and an ignition source are both present, yet no fire or explosion results. Because a vapor will explode or ignite if it can, there is clearly some condition present that prevents ignition. This saving factor is often the explosive limit of a flammable gas or vapor. In terms of our automobiles, a mixture of vapor and air below the lower flammable limit is too lean to burn. Conversely, a mixture of vapor and air above the upper flammable limit is too rich to burn. In the case of gasoline, concentrations of vapor below 1.4 percent and above 7.6 percent in air will not ignite in the presence of a significant ignition source. This spread between lower and upper limits is known as a liquid's **flammable range** and is specific to each individual flammable liquid or gas.

Specific Gravity

The weight of a solid or liquid substance as compared to the weight of an equal volume of water is its specific gravity. As in specific heat, the specific gravity of water is one. Any liquid or solid with a specific gravity of less than one will float on top of water; those with a specific gravity of greater than one will sink.

Vapor Density

The relative density of a vapor or gas is its vapor density. This is a ratio of any vapor or gas to that of air. Air, the standard, has a vapor density of one. Any vapor or gas with a vapor density less than one will rise; those with a vapor density greater than one will remain close to the ground.

OXYGEN

The third side of the fire tetrahedron, oxygen, is the most common element on earth. Oxygen makes up a major portion of the oceans and earth's crust, and one-fifth of our atmosphere. Atmospheric oxygen is the major concern of firefighters because it is the primary source of oxygen supporting the combustion cycle. Whereas other substances, such as chlorine and fluorine, can support combustion, most fires require and are aided by the presence of oxygen.

Oxygen itself does not burn. An electric arc or other spark will not explode any given concentration of oxygen. A match flame will ignite or explode a flammable gas, but the match itself burns faster when brought in contact with an oxygen-enriched atmosphere.

Oxygen supports combustion. Indeed, without oxygen, or some substitute, combustion is impossible. Normal burning is the combustion of fuels with oxygen under the influence of heat. However, it is important for the firefighter to remember that certain flammable chemicals have built into their molecules oxygen that is available for combustion even when there is no atmospheric oxygen available. Other materials, called oxidizing agents, release oxygen when heated. Although they may be nonflammable themselves, chemicals that release oxygen make any fire in nearby materials more intense.

Combustion

Rapid oxidation or chemical combination accompanied by the release of heat and light is called combustion. A heat source is applied to a piece of wood. The absorption of heat starts to decompose the wood, releasing particles of carbon and molecules of hydrogen and oxygen. These products of combustion mix with the air to form water, carbon monoxide, and heat energy in the form of a flame. The flame represents excess hydrogen gas and flammable carbon monoxide resulting from incomplete combustion. Unburned carbon particles are present in the smoke and charcoal ash remaining after combustion.

Oxidation versus Combustion

Oxidation is not necessarily combustion. Human beings take in about 200 gallons of atmospheric oxygen every day. We generate enough heat by the oxidation of fuels (food) to maintain our bodies at 98.6°F. Rust, the slow combination of iron with oxygen, is a form of oxidation. So are corrosion, the yellowing of the pages of old books and magazines, and certain explosions. Each of these represents a different rate of oxidation.

Spontaneous Ignition

Picture a pile of rags soaked in linseed oil. The linseed oil molecules are constantly being oxidized by the air. This combination releases energy. If the rags are tightly baled or stored in a tightly closed metal container, there is insufficient oxygen available for combustion. The formation of heat is held to a minimum. However, if the rags are loosely bound or piled in the open, exposed to air, there is sufficient air to allow the oxidation reaction to increase resulting in increased heat. The insulating effect of the pile of rags retards dissipation of the heat being generated by the chemical reaction. The temperature begins to rise, and the molecules become more energetic and combine with oxygen at an increasing rate. If left unchecked, this rise in temperature continues until the substance reaches its ignition temperature, resulting in spontaneous ignition.

Open Flames

When the heat of a burning match is applied to a piece of paper, the cellulose molecules of the paper start to vibrate at an enormously increased rate. Almost instantaneously they begin to break apart. In a series of reactions, these fragments continue to breakup, producing the free carbon and hydrogen necessary to combine with oxygen for combustion to take place. Once ignition occurs, the process continues to repeat itself until; (a) the entire fuel is consumed: (b) the fuel is cooled below its ignition temperature; (c) exposure to oxygen is mitigated; and/or (d) an interruption of the chemical chain reaction takes place to stop the process.

The assumption is that an ample supply of oxygen is available. It is important to remember that carbon is relatively slow burning, and normally there is insufficient oxygen available for complete combustion to occur. As a result, some carbon molecules leave the combustion zone as black smoke and floating particles. Some remain as charcoal, continuing to burn without flame. Still others form carbon monoxide, a toxic and flammable gas. Incomplete combustion, producing large amounts of carbon monoxide, is characteristic of gasoline engines and makes the inhalation of exhaust fumes dangerous, if not fatal.

Heat of Combustion

As discussed earlier, burning fuels produce heat energy. Because different fuels have a particular arrangement of atoms, each burns at a particular rate, different from the others. The amount of heat a fuel releases during complete oxidation is called its **heat of combustion.** Listed below are heats of combustion for some commonly encountered fuels, expressed in British Thermal Units (BTUs) released per pound of material:

Material	BTU/lb.
Paper, cardboard	6,000
Rags, wood	7,000
Newspaper, sawdust, fibers	8,000
Coal	12,500
Flammable liquids	16,000–21,000
Flammable gases	20,000–23,000
Hydrogen	60,000

Although these figures are only approximations, they give a good indication of the relative amounts of heat produced by various types of fuel.

Firefighters who are exposed to flammable liquid fires in the open air, such as vehicle fires involving gasoline, may underestimate their heats of combustion. The rate of heat rise in a flammable liquid or gas is much higher than that of wood or paper. Therefore, when a flammable liquid or gas is confined in a closed space, such as the interior of a building, the fire can double its heat in a relatively short period of time, making interior firefighting extremely difficult if not impossible. This is one reason why structure fires involving flammable liquids and gases present special problems.

STAGES OF COMBUSTION

Every combustion process involving ordinary combustibles in confined space goes through three distinct stages or phases. The change from one stage to another is caused by either an increase in heat production or a decrease in available oxygen. The transition from the first stage to the second results from increased heat production. The shift to stage three is the result of the consumption of available oxygen.

The **incipient phase,** the first stage of fire, is characterized by small flame production, minimal reduction of available oxygen, and low heat generation in the confined space. As the fire continues to burn, it moves into the second stage or **free-burning phase.** This phase is evidenced by free-burning in an oxygen-enriched atmosphere, heat production to a high of 1,300 degrees Fahrenheit, and an increase of heated gases rising into the upper atmosphere.

Continued combustion in the confined space ultimately results in a drastic reduction of available oxygen, forcing the fire into the third stage, the **smoldering phase.** This is the third, final, and most deadly phase of fire because the reduction of oxygen produces the largest amount of toxic fire gases, especially carbon monoxide. Burning is reduced to glowing embers, and there is an intense buildup of heat. The combustion process is stalled. Knowing what is taking place in the different stages of fire is necessary to gain an understanding of the resultant hazards to firefighters in a fire confined inside a structure.

HAZARDS OF COMBUSTION

The hazards of combustion are varied depending on the type of fire and fuels involved. Different fuels have different ignition temperatures and chemical composition. These variations make it difficult for firefighters to always predict what to expect. However, understanding of how fire burns and reacts during the different stages of fire enhances the ability of firefighters to forecast how a specific fire will burn. The following hazards have caused several fire deaths to both firefighters and civilians. Knowledge of the types of fires and in which stages these hazards are most likely to appear is a step toward reducing their adverse affect on life safety.

Flashover is a fireground phenomenon that has claimed the lives of many firefighters and is the result of the simultaneous ignition of all contents in a confined space. Flashover will most likely occur during the free burning phase of the fire. It is the result of the contents of a specified space reaching their ignition temperature as the result of the heated atmosphere. As the fire starts to burn, fire gases, lighter than air, begin to gather in the upper atmosphere. Continued free-burning increases the fire gases and heat in the upper atmosphere. Slowly all the contents, including the fire gases, are heated to their ignition temperature, and the entire contents explodes in flame. This flame spread is rapid and all consuming. It can take place in rooms other than the room on fire. There are incidents where flashover has occurred after initial knockdown of the fire. This is the result of hidden spaces holding heat and not sufficiently cooled by the initial application of water.

Backdraft occurs during the smoldering phase of the fire. It is characterized by the buildup of flammable fire gases that have been heated to their ignition temperature. However, they do not ignite because of insufficient available oxygen. The introduction of oxygen closes the fire tetrahedron, and ignition occurs. Again, ignition is explosive and destructive.

Fire gases vary greatly with the type of material burning. However, carbon monoxide, the most prevalent fire gas, is developed in almost every fire. Hydrogen cyanide, hydrogen sulfide, and hydrogen chloride are other gases that result from the burning of fuels containing those specific chemicals in their molecular structure. Regardless of the gas produced, fire gases can be separated into three classes; asphyxiants, irritants, or toxics.

Asphyxiants are gases that occlude oxygen from getting to the body's cells. This is accomplished either by displacing available oxygen in the air or through a chemical reaction that damages the hemoglobin's ability to carry oxygen to the cells. An example of an occluding-type asphyxiant is natural gas. Contrary to popular belief, breathing natural gas is not deadly. Death occurs once all the oxygen is displaced by the natural gas and the person dies from lack of oxygen. Once the victim is moved to an atmosphere containing oxygen, the oxygen exchange in the lungs will return to normal and the person can be resuscitated.

Carbon monoxide is an example of a fire gas classified as a chemical asphyxiant. Carbon monoxide is the most common chemical asphyxiant fire gas. Red blood cells have a greater affinity to carry carbon monoxide molecules than they do oxygen molecules. Therefore, when carbon monoxide is present in the lungs, the red blood cells choose them over any available oxygen. This process results in the inability of the red blood cells to carry oxygen even in the absence of additional carbon monoxide. That is why, even though victims may be removed to fresh air and given oxygen, rescuers may not be able to revive them.

Irritants are gases that cause irritation to the respiratory tract, causing difficulty in the exchange of oxygen in the lungs. Hydrogen chloride is an example of a fire gas in the irritant category. When inhaled, the chlorine mixes with saliva to create hydrochloric acid. The acid attacks the lining of the trachea and possibly the brachial tubes, causing fluid to collect in the lungs. Breathing becomes labored, and immediate medical attention is necessary.

Toxics are gases that actually attack internal organs or the nervous system. Most pesticides act on the nervous system, causing the nerves to misfire and malfunction. Most liquid pesticides are dissolved in a flammable solvent, and when heated, release toxic vapors much like nerve gas. Other toxics are known carcinogens and can cause cancer. Still others are capable of attacking the liver or kidneys, causing renal failure. The good news is that toxic fire gases are not present in every fire situation, and when they are, the quantities are minimal in concentration. The primary exception is the gases given off by burning plastics. The chemicals given off in the burning of most plastics are carcinogenic and have been known to cause cancer. Approach with caution.

FIRE EXTINGUISHMENT

Extinguishment of fire is a process of removing any one of the four sides of the fire tetrahedron. However, the diversity and complexity of the different chemicals and compounds makes selection of the appropriate method of fire extinguishment critical to a firefighter's success and survival. The basic accepted techniques of fire extinguishment are presented in this chapter. Specific extinguishing methods for different categories of chemicals are dependent upon the properties of those chemical compounds and will be discussed throughout this book.

Extinguishing Agents

An extinguishing agent is any substance applied to burning material that interrupts the combustion cycle. This interruption is accomplished by smothering the fire by excluding oxygen, cooling the burning material and/or atmosphere, separating the fuel from the heat and/or oxygen, or interrupting the chemical chain reaction. Extinguishing agents are chosen based on the physical properties of the burning material. The following are different types of extinguishing agents and how their application effects the combustion cycle.

Water. Its great absorption rate of heat required to turn it to steam facilitates the removal of heat from the burning process. However, the expansion rate of water to steam is 1,700 to 1. This expansion rate has a tendency to purge a confined area, removing oxygen, thereby providing a smothering effect on the fire. This expansion also purges most of the flammable gases created in the decomposition process, removing most of the fuel. Finally, because water is readily available, it can be used to channel fire in a direction more advantageous to the firefighter. An example would be the application of large streams of water to move burning flammable liquid away from tanks and buildings. Some flammable liquids have a specific gravity greater than water, and water will float on top of the liquid, excluding oxygen.

Every extinguishing agent has its negative factors. Water is no exception. In fact, the negative factors of water as an extinguishing agent are extreme in most situations where water is applied inappropriately. Water reacts violently with some materials, and there have been instances where firefighters have been killed merely by applying water to materials in an effort to extinguish fires. The specific gravity of water is such that many flammable liquids are lighter than it and will float on top of water. Some materials burn so hot that the application of water results in the liberation of hydrogen and oxygen molecules, greatly increasing the burning rate. Some applications of water result in water molecules becoming trapped beneath the surface of burning liquids. The water boils and the expansion of water to steam causes the burning material to spread in all directions, increasing fire spread. Finally, water is a very good conductor of electricity and has been known to travel down a fire stream, electrocuting the person holding the nozzle.

Carbon dioxide (CO_2) extinguishers. Carbon dioxide is heavier than air and when applied separates the oxygen from the burning material. Some cooling of the material takes place. However, the major effectiveness of this extinguishing agent is its smothering effect.

Dry chemical. Interrupts the chemical chain reaction. Application is directed at the base of the fire, where the chain reaction takes place.

Metal X. In addition to acting on the chemical chain reaction, metal x—a special form of dry chemical—forms a crust around the burning material in an effort to smother the fire. Metal x extinguishers are used to extinguish fire involving combustible metals, hence its name.

Because water is a firefighter's most plentiful resource, there is a tendency to rely on it in all situations. However, there are alternatives to consider when the application of water is inadvisable, unavailable, and/or ineffective. Some of these methods include:

• *Setting backfires:* Consume available fuel in a direction controlled by the firefighter.
• *Removing fuel:* Cut firebreaks, remove combustible materials ahead of the fire, ventilate to channel fire away from combustibles, drain flammable liquid tanks, and/or shut off flow.

Physical Properties and Extinguishment

As the population increases, so does the use of products made up of flammable liquids. This increased use and transportation of flammable liquid increase the opportunities for accidents. Flammable liquid incidents almost always result in major fires. This is primarily the result of bulk storage and transportation of flammable liquids. A single mistake can cause an incident that lasts days and/or results in significant loss of life. Flash point, specific gravity, ignition temperature, and flammable range all indicate difficulties in the extinguishment of flammable liquid fires. However, as much as these properties present significant difficulties to firefighters, it is in these same properties that the extinguishing methods are identified.

Water is firefighters' first line of defense. They are trained and equipped to think of using water first. However, there are situations in which the use of water is inadvisable and the use of other extinguishing agents is necessary. Therefore, anything that affects the usefulness of water should be of major concern to firefighters. Understanding the physical properties of flammable liquids helps identify the appropriate extinguishing method.

Use of Water on Flammable Liquids. Can you knock out a gasoline fire with a water fog? Although there are isolated incidents in which water fog has been effective in the extinguishment of gasoline-fed fires, the facts support a no answer. The important factors to consider are flash point and specific gravity. Extinguishment will occur when one of the sides of the fire tetrahedron is removed. Water extinguishes either by smothering the fire or by cooling it sufficiently to reduce the temperature. Because the flash point of gasoline is −46°F and that of hydrant water approximately 60°F it is highly unlikely that water application will result in the reduction of flammable vapors. Additionally, the specific gravity of gasoline is less than one, meaning that it will float on top of water. The application of water in an attempt to smother the fire will result only in spreading the gasoline and fire over a larger area.

Most authorities recommend the use of the proper types of foam or dry chemical on low flash point flammable liquids, with a possible supporting water–fog pattern to cool exposed surfaces. Water is much more effective on flammable liquids with a flash point between 100°F and 212°F, well above normal hydrant water temperature. In these cases, water is capable of cooling the liquid to below its flash point, stopping vapor production, and putting out the fire.

On combustible liquids with a flash point above 212°F, care must be taken not to allow water to become trapped below the surface of the burning liquid. Because of the flash point, the liquid itself is already above the boiling point of water. Water injected below the surface of the burning liquid will quickly turn to steam and increase 1,700 times its volume. A steam explosion will result, spitting hot liquid and spreading the fire. As the steam expands, the burning liquid can come in contact with other combustibles or with firefighters themselves, causing significant escalation of the incident. Excellent results have been obtained by the careful use of fog patterns that cool the liquid without allowing the water to become trapped beneath the surface.

The Dilution Technique.   Another physical property of flammable liquids is the ability (or not) to mix with water. Some liquids, such as acetone, ethyl alcohol, and methyl alcohol, are totally soluble (miscible) in water. This solubility is infinite, meaning that any portion of these liquids and water will mix.

Fire in a water-soluble liquid can be fought by entirely different methods than can fire in a liquid that is not water soluble. One such method is by dilution, adding water to the liquid, so as the percentage of water goes up in a water-soluble flammable liquid, so does the flash point. A 50 percent alcohol/water mixture has a flash point of 75°F (24°C); 10 percent alcohol in water has a flash point of 120°F (49°C).

In a spill fire, this dilution technique can be used to raise the flash point of a burning liquid so high that it will no longer produce flammable vapors. If the water-soluble liquid is not on fire, application of water may lower the temperature of the mixture below the level sufficient to produce potentially explosive vapors. This is another place where the high specific heat of water is of value to the firefighter.

The only problem in using the dilution technique involves the amount of water required. This varies from liquid to liquid. Ethyl alcohol requires the addition of from five to six times its volume of water to raise the flash point above 100°F (38°C). It would take the entire contents of a 300-gallon booster tank to dilute a 55-gallon drum of spilled alcohol. This amount of liquid should cause difficulties in containment and control of runoff, and many liquids require even more dilution. However, in a situation in which there are adequate supplies of water and no drainage problems, dilution is a fairly quick option to greatly reduce the hazard potential of the spill. Dilution is of less value in a flammable liquid tank fire. The required volume of water will quickly overflow the tank in most cases.

Using Foam. Because water has a higher specific gravity than most common flammable liquids, the effectiveness of water as an extinguishing agent on these types of fires is greatly reduced. To combat this disadvantage, scientists have developed a water additive that allows the water to float on top of the flammable liquid. The 3M Company was one of the first to market such an additive. Its brand name is AFFF or A triple F. The letters stand for Aqueous Film Forming Foam or "Light Water" because the product allows water to float on top of most flammable liquids.

If foam is chosen to fight a fire in a water-soluble liquid such as alcohol or acetone, it must be of a special type. Regular foam breaks down rapidly on contact with these liquids; a special "alcohol" foam should be used instead. However, in recent years, manufacturers of foam products have developed new foam concentrates that are effective on both hydrocarbon-based flammable liquid fires and alcohol fires. It is in firefighters' best interest to make sure which type of foam is carried on their apparatus and its proper application techniques.

Some liquids are only partially soluble in water. Many acetates and alcohols will mix only in small and variable amounts. The problem of what type of foam to use on "very slightly soluble" liquids is usually left unsolved. Most authorities simply recommend trying another type of foam if one does not work.

Fog Patterns. Water-soluble vapors and gases can sometimes be swept from the air with fog patterns. This offers an excellent way to reduce the explosive and toxic potentiality of the contaminated atmosphere around a fire. Spray patterns can also be used to dissipate insoluble gases or vapors by the force of the stream.

Even on flammable liquids that are not water soluble, hose streams can be used on spills to divert or control the spill in a preferred direction. This may be to direct the burning liquid away from or out from under flammable liquid tanks or to an area where the spill can be contained and foam applied to extinguish the fire. Remember: Whenever water is used on flammable liquid fires, extreme caution must be taken to predict and control runoff and possible escalation of the spill.

SUMMARY

Fire behavior is predictable and follows specific laws of physics and chemistry. The study of these laws will enable firefighters to analyze fire situations accurately, identify the specific hazards of materials involved, and develop appropriate action plans. The result will be more effective, efficient, and safer fireground operations.

REVIEW QUESTIONS

1. What are the four factors in the fire tetrahedron?
2. What are the four ways heat is generated?
3. Name five limitations of water as an extinguishing agent.
4. How is heat transferred?
5. Why is the latent heat of vaporization important to firefighting?
6. What is the importance of each of the following properties of flammable liquids:

   a. flash point
   b. ignition temperature
   c. flammable range
   d. water solubility

   e. specific gravity
   f. vapor density
   g. boiling point
   h. vapor pressure

FURTHER READING

1. Edwards, R.L.; *Fire Chem I. The Basics of H.T.M.* (4th ed.), S.A.F.E. Films, Inc.; 3326 Bentwood S.E., Grand Rapids, MI, 1990. Text supported with video-tapes.
2. *Essentials of Firefighting* (3rd. ed.), International Fire Service Training Association, Fire Protection Publications; Oklahoma State University, Stillwater, OK; 1994, pp. 5–30.

VISUAL AIDS

Applying Basic Chemistry. Chemistry of fire, vapor pressure, boiling point, flash point, fire point, ignition temperature, flammable/explosive limits, water reactive materials, pyrophoric and hypergolic materials, health hazards, and exposure limits. (34 minutes, VHS format; Media Resources, Inc.)

Flashover. Covers what a flashover is; how to assess its risks; what the danger signs are; survival techniques; how to avoid getting trapped during search and rescue. (VHS format; Fire Engineering Books & Videos, P.O. Box 21288, Tulsa, OK.)

Backdraft and Smoke Explosions. A combination of live fire scene footage, still photography, and graphic illustrations details what a backdraft is, how it occurs, what warning signs to look for, and what safety precautions to take. (20 minutes, VHS format; Fire Engineering Books and Videos, P.O. Box 21288, Tulsa, OK.)

DEMONSTRATIONS

Practice these demonstrations in an open area before class begins. Have several fairly large Pyrex beakers to contain the liquids and vapors and nonflammable pads for extinguishment. A small dry chemical extinguisher should be kept nearby to handle any mishaps. Remember that carbon disulfide, in particular, is toxic and has an extremely low ignition temperature. Use an extremely small quantity of carbon disulfide and do not allow its vapors to escape.

Four important properties can be demonstrated using four easily obtainable flammable liquids: gasoline, kerosene, carbon disulfide, and acetone.

1. *Flash Point* can be shown by holding a match over Pyrex beakers of gasoline and kerosene. When the gasoline ignites, the kerosene can be heated by the gasoline flame. Soon the kerosene can be ignited also, demonstrating the danger of exposure fires.

2. *Ignition temperature* can be shown with gasoline and carbon disulfide. (The heat of a burning cigarette probably will not ignite gasoline vapor, though it is possible for ignition to occur.) Therefore, the cigarette should be held in a gloved hand. The cigarette will immediately ignite the carbon disulfide.

3. *Specific gravity* can be demonstrated by the difference between gasoline and carbon disulfide. If water is added to the burning gasoline, the gasoline will float and will not be extinguished. Conversely, if water is added to the carbon disulfide, eventually the carbon disulfide will sink to the bottom, and the fire will be extinguished.

4. *Water solubility* can be shown by the gasoline already used and by acetone. Stir the beaker containing gasoline and water, and they will immediately separate. Ignite a small quantity of acetone in a beaker and extinguish it by covering with a nonflammable pad. Add enough water to dilute the acetone well, and it will not re-ignite. The dilution raises its flash point.

Chemical Properties of Hazardous Materials

Just as there are three physical states of matter—solids, liquids, and gases—there are three types of substances: elements, compounds, and mixtures. Elements and compounds are combined chemically, whereas mixtures can be made up of elements and/or compounds that have not chemically combined. Whether a substance is an element, compound, or mixture, its basic unit is an atom. **Atoms** are the "building blocks" of everything that occupies space and has weight. One hundred and six different atoms have been identified thus far. They are referred to as elements. An **element** is a pure substance that cannot be broken down into simpler substances by chemical means. Hydrogen and oxygen are two examples of elements.

ATOMS AND THEIR MAKEUP

Elements vary widely in their properties. Most are metals, and most are solids. Bromine, a nonmetal, and mercury, a metal, are liquids at room temperature. As far as we know, the entire universe is made up of fewer than one hundred of these elements, some of which are extremely rare. The other elements are artificially produced in the laboratory. Only eighty-eight occur naturally on our Earth, and only about thirty are at all common. Yet, these few "building blocks," combined in different arrangements, make up everything known to humanity. Elements combine with other elements based on specific laws of chemistry that are constant and predictable. An understanding of the laws of chemistry will enable firefighters to better predict possible outcomes of incidents involving hazardous chemicals. The increase in predictability could result in firefighters taking precautions that increase their level of safety.

Understanding the basic laws of chemistry requires an elementary understanding of the structure of an atom. The atom is divided into three basic parts: **protons, neutrons,** and **electrons**. Protons and neutrons are located in the nucleus of the atom. The nucleus is the center of the atom and contributes the major portion of the atom's mass. The electron resides in an orbit around the nucleus, much as the Earth orbits the sun. Like our solar system, the more electrons, the more orbits. Protons and electrons exhibit an electrical charge: Protons have a positive charge, electrons are negatively charged, and neutrons are neutral. Every atom maintains a neutral charge; therefore, the number of protons and electrons of an elemental atom are equal. The proton and neutron each have a weight of one unit, whereas the electron has negligible weight. Therefore, the **atomic weight** of the atom is a combination of the number of protons and neutrons.

Electron Shells

As stated earlier, electrons reside in orbits around the nucleus. The chemical nomenclature for these electron orbits is **shells.** For the purposes of understanding how elements combine to make compounds, fire service personnel need be concerned only with the outermost shell. Except for hydrogen and helium, all elements have the capacity to hold a maximum of eight electrons in their outer shell. Hydrogen and helium have a maximum capacity of two electrons in their outer shells. Everything in nature strives to be in its most stable condition. In the case of the atom, stability is achieved by maintaining eight electrons in the outer shell. This is known as the **octet rule**, or, in the case of hydrogen and helium, the **duet rule.** If an atom does not contain eight electrons in its outer shell, it is less stable than one that does. Instability can be resolved by letting go or collecting electrons until the outer shell is complete. When an atom loses or gains electrons, in an effort to become more stable the atom becomes an **ion**. Because electrons are electronically charged, once an atom gains or loses electrons, the ion becomes either positively or negatively charged. The loss or gain of electrons does not affect the atomic weight of the ion in that electrons have negligible weight.

Periodic Table

To better display the significance of electron shells, the elements have been arranged in a sequence called the **Periodic Table** (Table 2–1). The Periodic Table is a systematic arrangement of all the known elements by their **atomic numbers**. The atomic number represents the number of protons in the nucleus of the atom. The Periodic Table classifies the basic elements with respect to their relationship to one another and ultimately their hazards and potential lack of stability. The Periodic Table is made up of columns and rows. At the top of each column is a group number; for

TABLE 2-1 Periodic Table of Elements

Group Ia	Group IIa	Group IIIb	Group IVb	Group Vb	Group VIb	Group VIIb		Group VIIIb			Group Ib	Group IIb	Group IIIa	Group IVa	Group Va	Group VIa	Group VIIa	Group VIIIa
1 H Hydrogen																		2 He Helium
3 Li Lithium	4 Be Beryllium												5 B Boron	6 C Carbon	7 N Nitrogen	8 O Oxygen	9 F Fluorine	10 Ne Neon
11 Na Sodium	12 Mg Magnesium												13 Al Aluminum	14 Si Silicon	15 P Phosphorus	16 S Sulfur	17 Cl Chlorine	18 Ar Argon
19 K Potassium	20 Ca Calcium	21 Sc Scandium	22 Ti Titanium	23 V Vanadium	24 Cr Chromium	25 Mn Manganese	26 Fe Iron	27 Co Cobalt	28 Ni Nickel	29 Cu Copper	30 Zn Zinc	31 Ga Gallium	32 Ge Germanium	33 As Arsenic	34 Se Selenium	35 Br Bromine	36 Kr Krypton
37 Rb Rubidium	38 Sr Strontium	39 Y Yttrium	40 Zr Zirconium	41 Nb Niobium	42 Mo Molybdenum	43 Tc Technetium	44 Ru Ruthenium	45 Rh Rhodium	46 Pd Palladium	47 Ag Silver	48 Cd Cadmium	49 In Indium	50 Sn Tin	51 Sb Antimony	52 Te Tellurium	53 I Iodine	54 Xe Xenon
55 Cs Cesium	56 Ba Barium	57 La Lanthanum	72 Hf Hafnium	73 Ta Tantalum	74 W Tungsten	75 Re Rhenium	76 Os Osmium	77 Ir Iridium	78 Pt Platinum	79 Au Gold	80 Hg Mercury	81 Tl Thallium	82 Pb Lead	83 Bi Bismuth	84 Po Polonium	85 At Astatine	86 Rn Radon
87 Fr Francium	88 Ra Radium	89 Ac Actinium															

Rare Earth Metals

58 Ce Cerium	59 Pr Praseodymium	60 Nd Neodymium	61 Pm Promethium	62 Sm Samarium	63 Eu Europium	64 Gd Gadolinium	65 Tb Terbium	66 Dy Dysprosium	67 Ho Holmium	68 Er Erbium	69 Tm Thulium	70 Yb Ytterbium	71 Lu Letetium

Actinide Series

90 Th Thorium	91 Pa Protactinium	92 U Uranium	93 Np Neptunium	94 Pu Plutonium	95 Am Americium	96 Cm Curium	97 Bk Berkelium	98 Cf Californium	99 Es Einsteinium	100 Fm Fermium	101 Md Mendelevium	102 No Nobelium	103 Lw Lawrencium

example, the first column of elements on the left is labeled "Group Ia." "Group a" elements have similar characteristics, and their behavior is predictable. "Group b" elements (located in the center of the table), the Rare Earth Metals, and the Actinide Series (both located at the bottom of the table) are very complex in structure and do not exhibit predictable behavior. Therefore, our discussion will focus on understanding the characteristics of Group a elements and how elements chemically bond to form compounds.

BONDING AND THE FORMATION OF MOLECULES AND COMPOUNDS

Valence

The numeric of the Group a number depicts the number of electrons in the outer shell. Elements in Group Ia have one electron in their outer shell. Conversely, elements in Group VIIa have seven electrons in their outer shell. Group VIIIa is special—elements in this group have all eight electrons in their outer shell and are referred to as **noble gases.** These elements are said to be inert, meaning nonreactive. Typically, they are found by themselves and do not tend to combine with other elements. Other elements in Group a strive for this same level of stability. To accomplish this, an atom will either release or collect electrons to look like their closest noble gas counterpart. Lithium, a Group Ia element, will release an electron to look more like helium. Likewise, fluorine, a Group VIIa element, wants to gather one more electron to look more like neon. When the electrons are released or gathered, the resultant ion takes on either a negative or positive charge. This charge is called **valence.** If the element receives or releases an electron it will exhibit a negative or positive valence, respectively. The amount of that valence is determined by the group number. Elements in Group Ia will have a valence of +1; elements in Group IIa, +2; Group VIa, –2, and so on.

Chemical Bonding

The valence of an atom determines how that element will combine with other atoms. There are two different ways valence works toward chemically bonding two or more elements together. One type of bonding occurs when the electrical charge of an ion acts as a magnet looking for an oppositely charged particle to gain neutrality and stability. This type of bonding is called ionic bonding. The second involves the sharing of electrons. This happens when two or more elements combine by sharing their electrons to fill up their outer shell. Two hydrogen atoms, with one electron each, share both with oxygen's two vacancies in its outer shell. The result is called **covalent bonding**. Covalent bonding may result in the sharing of one, two, three, or four pairs of electrons, each atom acting as if all the shared electrons belong to them.

Metals and Nonmetals

A further distinction evident in the Periodic Table is the separation of metals and nonmetals. This separation is identified by a bold line between the metals and nonmetals. This line starts to the left of boron and extends diagonally to between polonium and astatine. Hydrogen is also a nonmetal, however. Because its chemical bonding characteristics most accurately align with those of Group Ia metals, it appears in that column in most Periodic Tables. The relationship between metals and nonmetals is much more significant than just their location on the Periodic Table. Compounds composed of metals and nonmetals are always the result of ionic bonding and are referred to as **salts**. Sodium chloride (table salt) is an example. Compounds composed of two or more nonmetals contain covalent bonding and are referred to as **nonsalts**. Carbon dioxide (CO_2) is an example. Ions of metal elements always have a positive valence. Ions of nonmetal elements, in salt compounds, always carry a negative valence, with the exception of hydrogen. Compounds between nonmetals are dependent on the sharing of electrons and may not follow the same rules as ionic bonding.

A **molecule** is a single unit of two or more atoms that have chemically combined, regardless of whether the atoms are the same, nonmetals, or metals and nonmetals. Hydrogen, nitrogen, oxygen, fluorine, chlorine, bromine, and iodine combine with themselves to form molecules that are called **diatomic**.

ATOMIC SYMBOLS

In order to show what elements or combinations of elements are present in a substance, it is both convenient and necessary to learn a sort of chemical shorthand. Each element is represented by either a single letter or no more than two letters. C stands for carbon; H for hydrogen; O, oxygen; P, phosphorus. Two-letter symbols include Cl, chlorine; He, helium; Li, lithium; Ne, neon. When one letter is used, it is always capitalized. When two letters are used, the first letter is capitalized but the second letter is not.

Some symbols of elements do not seem to match their names at all. Some of the symbols, for example, come from ancient Latin names: Fe, iron, from *ferrum*; Au, gold, from *aurum*; Pb, lead, from *plumbum* (from which the word *plumber* is derived); and K, potassium, from *kalium*.

Familiarity with these words will come from repeated use, just as *lb* is automatically substituted for pound. Now is a good time to take a moment and study the Periodic Table to gain familiarity with the different elements, their names, and their relationship to other elements in the table.

Radical Ions

Up to this point the discussion has focused on the elements. However, compounds are made up of more than just two elements. If there are more than two elements chemically combined in a compound, how does that follow the bonding rules just discussed? In the case of covalent bonding, the answer is easy—more elements share more electrons. However, how are metals and more than one nonmetal element combined, like sodium hydroxide, NaOH? The answer is **radical ions**. Radical ions are the result of two or more atoms' covalently bonding together yet remaining negatively charged. As ions, they are then capable of ionically bonding with metal ions. There is no set formula for determining the valences of these radical ions; therefore, it is easiest to simply commit them to memory. The most-used radical ions and their respective valences are depicted as follows:

Sulfate	$(SO_4)^{-2}$	Sulfite	$(SO_3)^{-2}$
Nitrate	$(NO_3)^{-1}$	Nitrite	$(NO_2)^{-1}$
Carbonate	$(CO_3)^{-2}$	Bicarbonate	$(HCO_3)^{-1}$
Acetate	$(C_2H_3O_2)^{-1}$	Borate	$(BO_4)^{-2}$
Chromate	$(CrO_4)^{-2}$	Ammonium	$(NH_4)^{+1}$

Notice that the radicals ending in "ate" or "ite" contain oxygen in their formulae. This is a clear indicator to firefighters that these types of compounds, in fire, will produce oxygen and may increase fire intensity. Here is a graphic example of how a little knowledge of basic chemistry may give firefighters the upper hand in protecting life and property.

Electron Dot Symbols

Thus far, elements have been represented by their elemental symbol. Prior to the discussion of equations, there is a more understandable representation of the elements called electron dots. All the electron dot symbol does is combine the atomic symbol with the electron configuration of its outer shell. Examples of electron dot symbols are shown in Figure 2–1.

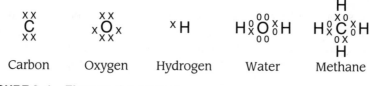

| Carbon | Oxygen | Hydrogen | Water | Methane |

FIGURE 2–1   Electron dot symbols.

This subtle change in the basic symbol makes it easier to explain how elements combine to form compounds. As long as there are dots left over, more elements are capable of attaching. Another way of determining the structure of a molecule is to look at its structural formula. Butane, classified as a liquid petroleum gas (LPG), has a molecular formula of C_4H_{10}, meaning that 4 carbon atoms link with 10 hydrogen atoms to form a molecule of butane. From the electron dot formula of carbon, we can see that there are four vacant positions for each carbon atom. Hydrogen has one vacant position per atom. Therefore each carbon atom will share one electron with each hydrogen atom and one or more additional carbon atoms. This is accomplished by the first carbon atom's sharing three vacant positions with three hydrogen atoms and one vacant position with an additional carbon atom. The second carbon atom shares one electron with the first carbon atom, a second electron with the third carbon atom, and the remaining two electrons with two hydrogen atoms. The third carbon atom is configured like the second, and the fourth is similar to the first. This accounts for all four carbon atoms and ten hydrogen atoms. Figure 2–2 depicts the electron dot formula for butane by replacing the dots with a bar to depict a single covalent bond. Double and triple covalent bonds also exist, and their significance will be discussed in Chapter 3.

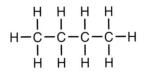

Butane

FIGURE 2–2

Process Symbols

Often, in describing the way two different compounds combine, arrows show the way the process proceeds. An arrow pointing upwards (↑) shows formation of a gas. An arrow pointing downward (↓) symbolizes formation of a solid, often called a precipitate. Used at the center, an arrow points out the direction of the reaction (→). Arrows pointing in opposite directions (←→) show that the process is reversible.

CONSTRUCTING AN EQUATION

The next step, once an understanding of chemical symboling is mastered, is to learn how chemicals are combined to form compounds. Common table salt, chemical name sodium chloride, has a molecular structure of NaCl. The chemical equation that depicts the formation of sodium chloride is Na+Cl → NaCl. In lay terms, one atom of sodium combines with one atom of chlorine to form sodium chloride. The formation of water can be depicted in the same manner. However, this time the molecular formula, H_2O, states that it takes two hydrogen atoms for every one oxygen atom. Couple this with the fact that both hydrogen and oxygen are diatomic (reside only as two atoms) in nature, and the equation looks like this $2H_2+O_2 \rightarrow 2(H_2O)$. Two molecules of hydrogen combine with one molecule of oxygen to form two molecules of water.

Notice that all the atoms are accounted for on both sides of the equations. There are the same number of atoms of each element on each side of the center arrow. Remember: On the molecular level, whether in a reaction of this kind or in combustion, every atom remains the same, although it may shift position and join with one or more other atoms to form a new compound or molecule.

The chemical representation for the simple combustion of carbon and oxygen can now be represented in chemical formula. A molecule of elemental oxygen contains two oxygen atoms, O_2. When a carbon atom combines with oxygen, it forms a combustion product, carbon dioxide, $C+O_2 \rightarrow CO_2$, and energy is released in the form of heat. However, if there is insufficient oxygen present, two carbon atoms combine with one molecule of oxygen to form two molecules of carbon monoxide, $2C+O_2 \rightarrow 2CO$. Again, energy is released in the form of heat.

In Chapter 1, the reaction chain discussed relative to open flames and burning wood can now be equated chemically to show all the by-products of combustion.

$$C_6H_{10}O_5 + 6O_2 \rightarrow 5H_2O+CO_2\uparrow \text{ and heat energy}$$

As the rapid burning increases and/or the oxygen diminishes, unburned carbon particles in the form of black smoke become apparent, and carbon monoxide is produced. Because heat is liberated, the process continues until all fuel is consumed. The only other factors to affect change in this process would be cooling of the burning material, elimination of the oxygen, or insertion of chemicals to break up the chemical chain reaction. These are commonly known as extinguishing methods.

VALENCE, BONDING, STABILITY, AND FLAMMABILITY

As stated earlier, the valence of any atom determines how it will combine with other atoms. It was determined that the carbon atom had a valence of four and would covalently combine with other atoms in sufficient quantity to reach a combined total valence of eight. One carbon atom will combine with four hydrogen atoms to form molecules of methane, the primary ingredient of natural gas, or with four chlorine atoms, to form molecules of carbon tetrachloride. Similarly, nitrogen, generally having a valence of three, will combine with three hydrogen atoms to form a molecule of ammonia.

Carbon has a valence of four; oxygen, a valence of two. When carbon and oxygen combine, each oxygen atom satisfies two of the carbon valences. This results in two single bonds, and carbon dioxide molecules are formed. When this happens, all the valences have been satisfied, and the carbon atom has been fully oxidized. It cannot combine with additional oxygen; it is completely "burned" already. This is why CO_2 is generally inert and nonflammable. However, a carbon monoxide molecule is the result of an oxygen-deficient atmosphere and has not reached its satisfied state of complete oxidation. Two valences are left unused, resulting in the **double bonding** of carbon and oxygen. The double bonding results in a less stable molecule, and carbon is still capable of combining with other elements. Therefore, it is highly toxic and flammable, contributing to flame spread and fire deaths.

During the discussion on ionic bonding, the statement was made that nonsalt compounds did not always follow set rules. In the case of the carbon atom, it is capable of single, double, and even triple bonds. Compounds that are the results of double and triple bonds are less stable and tend to be more dangerous than those of single bond unions. The significance and impact of this phenomenon will be discussed, in depth, in Chapter 3.

ORGANIC COMPOUNDS

Organic chemistry is the study of compounds that contain carbon in their molecular structure. They exist in all three physical states of matter. Organic solids are classified into two categories, hydrocarbon based and cellulose based. Hydrocarbon-based solids are synthetic and consist primarily of plastic compounds. Cellulose-based materials include wood, paper, and cotton and are naturally made. This book focuses on flammable hazardous materials. Cellulose-based organic solids are characterized by ordinary combustibles and will not be discussed further. However, hydrocarbon-based solids and carbon-based liquids have significant hazards associated with their combustion. Plastics are discussed in Chapter 10. The discussion on flammable liquids will talk place in Chapter 3.

SUMMARY

Although it is highly unlikely that the average firefighter will become an expert in the area of chemical composition and reactions, an understanding of the basic laws of chemistry will save lives. The ability to identify common compounds and forecast the possible outcomes of mixing those compounds or how they will react in a fire situation is critical to firefighting personnel on-scene. How chemicals bond, whether they are stable or unstable, if they are reactive in a given situation, and what effects extinguishing agents might have on them is all answerable through a basic knowledge of chemistry. Understanding atomic structure, the concept of ionic and covalent bonding, aromatic ringed hydrocarbons, and organic chemistry will increase the successes at the scene of an emergency. Knowing the characteristics of a specific hazardous material will enable the firefighter to develop the most effective and efficient strategies and tactics to address any situation. It all begins with a basic understanding of chemistry.

REVIEW QUESTIONS

1. Define an element, a compound, and a mixture.
2. Is H_2O a molecular formula or a structural formula?
3. Why won't carbon dioxide burn?
4. What dangerous gas is produced when carbon combines with oxygen in an enclosed space?
5. What is the definition of an inert gas?
6. What is the difference between ionic bonding and covalent bonding?
7. What is the significance of single, double, and triple bonds?
8. What is organic chemistry?
9. What is the "octet rule," and how does it affect chemical reactions?
10. What is the difference between a salt and a nonsalt?
11. What is a radical?
12. What is meant by the term *diatomic*?
13. Write the chemical equation for the formation of the following compounds:
    a. sodium and chlorine to make sodium chloride
    b. carbon and oxygen to make carbon dioxide
    c. carbon and oxygen to make carbon monoxide
    d. carbon and hydrogen to make butane
    e. nitrogen and hydrogen to make ammonia

FURTHER READING

1. Edwards, R.L. *Fire Chem I.The Basics of H.T.M.,* (4th ed.), pp. 27-68. Films, Inc.; 3326 Bentwood S.E., Grand Rapids, MI. 1990. Text supported with videotapes.

Flammable Liquids

3

The primary reason flammable liquids cause or are involved in so many fires, and cause such great loss of life is that they are so common and plentiful in today's society. We are surrounded by a sea of liquids that can burn. As fuels, flammable liquids power everything from lawnmowers to jet airplanes. As solvents, flammable liquids can be found in an amazing variety of industrial products, from paints to insecticides, inks to adhesives. Although there is an environmental push to reduce the proliferation of hydrocarbon-based products, make no mistake: more exotic, toxic, and flammable products are being produced today than ever before.

The National Fire Protection Association (NFPA) defines a flammable liquid as one with a flash point below 100°F (38°C) and a vapor pressure not exceeding 40 pounds per square inch. A combustible liquid is any liquid whose flash point is 100°F (38°C) or above. This information is of great importance to firefighters in that it allows them to determine the likelihood of a liquid's igniting given its temperature and flash point. Flammable liquids and combustible liquids are subdivided into classes, as shown in Tables 3–1 and 3–2.

TABLE 3–1. Classes of Flammable Liquids

Class	Flash Point	Boiling Point
IA	Below 73°F (23°C)	Below 100°F (38°C)
IB	Below 73°F (23°C)	At or above 100°F (38°C)
IC	At or above 73°F (23°C) and below 100°F (38°C)	Below 100°F (38°C)

29

TABLE 3–2. Classes of Combustible Liquids

Class	Flash Point
II	At or above 73°F (23°C) and below 140°F (60°C)
IIIA	At or above 140°F (60°C) and below 200°F (93°C)
IIIB	At or above 200°F (93°C)

In previous chapters, the properties of different types of fuels have been discussed, including those shared by all flammable liquids. Table 3–3 lists eight common flammable liquids, each of which also represents a much larger class. Acetone is an example of a ketone. Isoamyl acetate is an ester. Also included are an ether, an alcohol, an amine, an aldehyde, and a hydrocarbon. This chapter will discuss how these classes of liquids differ from one another in uses and hazards. Although there are thousands of different flammable liquids, the number of physical properties that are important to the safety services are few. Failure to understand the properties of a liquid implies lack of awareness of its potential danger and behavior.

The headings in Table 3–3 were discussed in Chapter 1. Their relationship to one another and importance to firefighters is critical for the effective handling of an incident involving flammable liquids. Firefighters who know how to interpret the table will be capable of predicting how a particular liquid will: (1) act in a fire; (2) under what conditions, ignite or explode; and, (3) ultimately, how it should be handled in an emergency. The seven key terms appearing above the table columns explain the hazards of thousands of liquids that may differ completely in their uses. However, though they may differ greatly in their uses, commonalty of the seven key factors dictates the correct action to take in an emergency.

HYDROCARBONS

Crude oil pumped from the ground is a mixture of many different molecules. It is refined into usable products (called fractions) by an industrial process called fractional distillation. The various molecules are separated, collected at different levels of a fractioning tower, piped away, cooled, and blended into familiar products (see Table 3–4).

Gasoline, kerosene, naphtha, petroleum ether, fuel oil, lubricating oils, asphalt, and other liquid petroleum fractions are all hydrocarbons. Their molecules are made up of hydrogen and carbon. Carbon atoms can link together to form long chains containing hundreds or even thousands of atoms. However, this chapter is concerned with the shorter chains that form flammable liquids.

TABLE 3–3. Properties of Some Common Flammable Liquids

	Flash Point	Ignition Temp	Lower Flame Limit	Upper Flame Limit	Specific Gravity	Vapor Density	Boiling Point	Water Soluble
UL Class 90: Acetone (dimethyl ketone) CH_3COCH_3	0°F 18°C	1,000°F 538°C	2.6	12.8	0.8	2.0	134°F 57°C	Yes
Acetaldehyde (ethanol or acetic aldehyde) CH_3CHO	−36°F −38°C	365°F 185°C	4.1	55.0	0.8	1.5	70°F 21°C	Yes
UL Class 55–60: Amyl acetate–iso (banana oil) $CH_3COOCH_2CH(CH_3)_2$	77°F 25°C	715°F 379°C	1.0	7.5	0.9	4.5	290°F 143°C	Slightly
UL Class 110 plus: Carbon disulfide (Carbon bisulfide) CS_2	−22°F −30°C	212°F 100°C	1.3	44.0	1.3	2.6	115°F 46°C	No
UL Class 70: Ethyl alcohol (ethanol or grain alcohol) C_2H_5	55°F[a] 13°C	793°F 423°C	4.3	19.0	0.8	1.6	173°F 78°C	Yes
Ethyl amine (amino ethane) 70% aqueous solution $C_2H_5NH_2$	<0°F <−18°C	723°F 384°C	3.5	14.0	0.8	1.6	62°F 17°C	Yes
ULClass 100: Ethyl ether (Diethyl ether or ethyl oxide) $C_2H_5OC_2H_5$	−49°F −45°C	356°F 180°C	1.9	48.0	0.7	2.6	95°F 35°C	Slightly
Gasoline C_5H_{12} to C_9H_{20}	−36°F[b] −38°F[b]	536°F[b] 280°C	1.4	7.4	0.8	3–4	100°F[b] 38°C[b]	No

[a] As ethyl alcohol is mixed with water, the flash point changes. Fifty percent alcohol and 50 percent water has a flash point of 75°F (24°C). Ten percent alcohol and 90 percent water has a flash point of 120°F (49°C).
[b] The properties of gasoline vary according to the octane.

Table 3–4. Properties of Typical Petroleum Fractions

	Flash Point	Ignition Temp.	Lower Flammable Limits	Upper Flammable Limits	Specific Gravity	Vapor Density	Boiling Point
Asphalt (petroleum pitch)	400°F[a] 204°C	905°F 485°C			1.0–1.1		700°F 371°C
Crude petroleum	20°–90°F -7°–32°C				1.0		
Fuel oil #1 (kerosene)	100°F[b] 38°C[b]	444°F 229°C	0.7	5.0	≥1.0		304°–574°F 151°–301°C
Fuel oil #2	100°F[b] 38°C[b]	494°F 257°C			≥1.0		700°F 371°C
Fuel oil #4	130°F[b] 54°C[b]	505°F 263°C			≥1.0		700°F 371°C
Fuel oil #5	130°F 54°C[b]				≥1.0		700°F 371°C
Fuel oil #6	150°F[b] 66°C[b]	765°F 407°C			±1.0		700°F 371°C
Gasoline: 60 octane	-45°F -43°C	536°F 280°C	1.4	7.6	0.8		100°–400°F
100 octane	-36°F -38°C	853°F 456°C				3.0–4.0	38°–204°C
Lubricating oil (motor oil)	300°–450°F 149°–232°C	500°–700°F 260°–371°F			≥1.0		680°F 360°C
Mineral oil	380°F 193°C				0.8–0.9		680°F 360°C
Naphtha—Safety solvent (Stoddard solvent)	100°–140°F 38°–60°C	300°–450°F 149°–232°C	1.1	6.0	0.8		≤357°F ≤181°C
Petroleum ether (benzin, ligroin)	-50°–0°F -46°–18°C	550°F 228°C	1.1	5.9	0.6	2.5	95°–140°F 35°–60°C

[a] Depending on the curing qualities, the flash point of asphalt ranges from 100°F to 250°F (38°C to 121°C). Cutback asphalt has a flash point of 50°F (10°C).
[b] Minimum or legal flash point, which varies in different states.

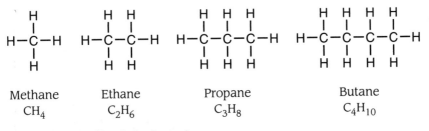

FIGURE 3–1. Simple hydrocarbons.

The one significant similarity of the hydrocarbons is the methane radical as their foundation. The methane radical is the methane molecule minus one hydrogen atom. The open bond is left for other radicals or atoms to attach to, forming the different classes of flammable hydrocarbons. The atom or radical added to the basic hydrocarbon structure is identified by the molecular formula in Figure 3–1.

Alkanes

The basic methane molecule identified in Chapter 2 is the basis for the alkanes. Each additional carbon atom replaces one hydrogen atom to continue the chain. Alkanes are the least complex set of hydrocarbons. Characterized by the "ane" suffix in the compound name, they are commonly referred to as the **paraffin series**. The prefixes are derived from Greek and represent the number of carbon atoms in the named compound. *Meth* equals one carbon atom; *eth*, two; *prop*, three; *but*, four; *pent*, five; and so on. Notice in Figure 3–1 that paraffin chains can be built indefinitely. Each molecule contains twice as many hydrogen atoms, plus two, as carbon atoms. All alkanes are **saturated hydrocarbons**, meaning that all available carbon bonds are used up in a single bond configuration. Other hydrocarbon series are **unsaturated** and will be discussed later. Table 3–5 lists the members of the paraffin series up to a ten-carbon chain. Most of the fire incidents relative to the paraffin series are caused by the flammable gases and liquids of the smaller chained hydrocarbons. However, the following incident involving paraffin wax is a reminder that all compounds in the paraffin series can cause significant damage when involved in fire.

> On December 8, 1992 at 3:15 A.M., firefighters were summoned to a fire in a single story wax processing plant that processed raw paraffin wax into food grade wax for food processing. The first company attempted an offensive attack on the interior with an inch and three-quarter handline. Interior crews stated that the fire was so intense and its spread was so rapid, they were only able to penetrate approximately 30 feet into the structure. Interior crews were pulled out and within nine minutes, the roof collapsed. Firefighters on-scene stated that, "Once the wax melted at approximately 125 degrees Fahrenheit, it acted just like a 'free-flowing' fuel oil fire." Heavy streams were used to cool

TABLE 3-5. Paraffin Series

	Flash Point	Ignition Temp.	Lower Flammable Limits	Upper Flammable Limits	Specific Gravity	Vapor Density	Boiling Point	Remarks
Methane CH_4	Gas	999°F 537°C	5.3	14.0		0.6	−259°F −162°C	Colorless, odorless, asphyxiating gas. Chief component of natural gas.
Ethane C_2H_6	Gas	959°F 515°C	3.0	12.5		1.0	−128°F −89°C	Colorless, odorless, asphyxiating gas. Used in the manufacturing of organic compounds.
Propane C_3H_8	Gas	871°F 466°C	2.2	9.5		1.6	−44°F −42°C	Colorless, asphyxiating gas with little natural odor. One of the LPGs.
Butane C_4H_{10}	Gas	761°F 405°C	1.9	8.5		2.0	31°F −0.6°C	Colorless, asphyxiating gas with little natural odor. One of the LPGs.
Pentane C_5H_{12}	−40°F[a] −40°C[a]	588°F 309°C	1.5	7.8	0.6	2.5	97°F 37°C	Colorless liquid with a pleasant odor. Used as a solvent, anesthetic, and in artificial ice production.
Hexane C_6H_{14}	−7°F −28°C	453°F 234°C	1.2	7.5	0.7	3.0	156°F 69°C	Colorless liquid with a faint odor. Used as a solvent and as a liquid in low-temp thermometers.
Heptane C_7H_{16}	25°F −4°C	433°F 223°C	1.2	6.7	0.7	3.5	209°F 98°C	Colorless liquid used as a solvent and anesthetic. One of the primary ingredients of gasoline.
Octane C_8H_{18}	56°F 13°C	588°F 309°C	1.0	3.2	0.7	3.9	258°F 126°C	Colorless liquid used as a solvent and in the manufacture of chemicals. A standard for gasoline rating.
Nonane C_9H_{20}	88°F 31°C	403°F 206°C	0.8	2.9	0.7	4.4	303°F 151°C	Colorless liquid used in the manufacture of organic chemicals; gasoline, kerosene, and naphtha.
Decane $C_{10}H_{22}$	115°F 46°C	406°F 208°C	0.8	5.4	0.7	4.9	345°F 174°C	Colorless liquid; principle component of kerosene and naphtha.

[a]Minus 40°F and minus 40°C are the only points on the temperature scale where the readings are identical.

and solidify the wax. However, ordinary combustibles involved in the fire continued to maintain the liquid consistency of the wax. Once the ordinary combustibles were consumed, the wax could be cooled below its melting point and extinguishment was attainable.

As the molecules in a particular series became larger, they naturally become heavier. Methane and ethane are gases. Propane and butane are gases that can be liquefied easily by pressure; they are the LP (liquid petroleum) gases. Butane, a gas at room temperature, will liquefy on a cold day. When the chain becomes five carbon atoms long, the molecules are too large to be gases at ordinary temperatures. Pentane is a liquid, as are hexane, heptane, octane, nonane, and decane. Eventually the carbon chain becomes so long that semisolids and solids are formed. Petroleum jelly (Vaseline), paraffin wax, and asphalt are examples.

Gasoline

Gasoline is a typical liquid hydrocarbon. It burns vigorously with a bright orange flame and emits large quantities of black smoke. It is lighter than water and not miscible in it. Although not considered highly toxic, gasoline can quickly build up lethal concentrations of carbon monoxide when burning in an enclosed space, in or out of an internal combustion engine. It is also a skin irritant, and prolonged inhalation of the vapors in a confined space has been known to be fatal.

Structurally, gasoline is a blend of isomers. Sometimes, the same number of atoms can hook together in a slightly different manner. This revised lineup is called an **isomer**. Octane and iso-octane have the same molecular formula: C_8H_{18}. However, structurally they are not exactly the same (see Figure 3–2).

Modern gasolines are blended from a wide variety of hydrocarbons (and their isomers) ranging from pentane to nonane. Occasionally, benzene, toluene, and certain alcohols are included in the blends. Regardless of the different blends, the importance for firefighters remains gasoline's ease of ignition and difficulties in extinguishment.

Misuse of gasoline constitutes one of the major fire hazards in our country today. Of necessity, gasoline is universally available. In addition to

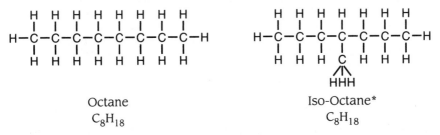

Octane
C_8H_{18}

Iso-Octane*
C_8H_{18}

FIGURE 3–2. Isomers found in gasoline.
* Molecules such as iso-octane are called branched hydrocarbons.

being a fuel, gasoline is used as a solvent in certain industrial processes. As a consequence of its familiarity, it is treated with a measure of complacency, but rest assured, it is still an extremely dangerous, flammable liquid. It is ready to burn at any time. The misuse of gasoline for washing automotive parts or for dry cleaning creates an explosive air mixture that requires only a source of ignition. Such was an incident in the city of Ontario in the mid-1970s. A man had just finished overhauling his engine. While cleaning up, he washed his hands and arms with gasoline and entered the house to take a shower. He got as far as the water heater when the pilot light ignited the gasoline vapors, resulting in second and third degree burns over a significant portion of his body.

Educating the public about misuses of gasoline is a major concern of the fire service. Misrepresentations by movie makers do not help in the development of that education. Numerous films depict flammable liquid hazards in a less than factual manner. Contrary to popular belief, most cars do not explode on impact, and the fuel used in commercial jets will not ignite when spilled on the ground in a blinding snowstorm. Conversely, a scene showing two individuals tossing lighted matches at each other's gasoline-soaked clothing without ignition is also a tragedy. It leaves the audience with the perception that as long as contact with the clothing is avoided, ignition will not take place. However, ignition of the vapors is what causes gasoline to burn. All that is needed is the correct vapor–air mixture.

The following is an excerpt from a National Transportation Safety Board investigation of an incident involving gasoline:

At about 12:12A.M., on April 7, 1982, several vehicles on westbound California State Route 24 entered the north, No. 3 Bore of the Caldecott Tunnel near Oakland, California. A Honda car driven by an intoxicated driver struck the raised curbs inside the tunnel and came to rest at the left edge of the roadway about one-third of the way through the tunnel. It was struck soon afterward by a gasoline tank truck and tank trailer and then by an AC Transit bus, which subsequently struck the tank trailer. The bus driver was ejected and the empty bus continued west, exited the tunnel, and struck a concrete road support pier. The tank trailer overturned and gasoline was spilled inside the tunnel. A fire erupted and heavy black smoke quickly filled the tunnel. The tank truck and tank trailer, the Honda car, and four other vehicles that had entered the tunnel were completely destroyed by the fire. Seven persons were killed and two people were treated for minor smoke inhalation. The tunnel incurred major damage.

Of the seven fatalities, one was contributed to accident trauma, the other six to smoke inhalation and burns. The carbon monoxide level in the six victims ranged from a low of 4.1 percent to a high of 31 percent. Analysis of the incident revealed that the temperatures reached in the tunnel exceeded 1,500°F at the roadbed.

Kerosene (Kerosene, Range Oil, Fuel Oil #1, Coal Oil)

Kerosene is a mixture of hydrocarbons having ten to sixteen carbon atoms in their molecules (decane to hexadecane). It is a pale yellow to white liquid, with a strong, familiar odor. Skin-irritating qualities are about the same as those of gasoline and, as with all hydrocarbons, carbon monoxide can be a product of incomplete combustion. Kerosene is formulated to have a flash point between 100°F (38°C) and 140°F (60°C). With a flash point in this range, it will not produce flammable vapors at ordinary temperatures. Used as a cleaner or degreaser, it is much safer than gasoline. However, if kerosene is heated to its flash point, the vapors are readily ignited, and its heat of combustion is high. Kerosene burns hot, and its fire potential when exposed to heat should not be underestimated.

Kerosene was once the most economically important fraction of petroleum. Homes were lit by kerosene lamps and heated by coal oil stoves. Although electricity and natural gas have greatly reduced the use of kerosene for these purposes, portable kerosene heaters are still a major cause of loss of life by fire and asphyxiation.

Kerosene is also widely used as a cleaner, degreaser, and solvent. In recent years, it has staged a comeback as a fuel. Military and commercial jet aircraft generally burn kerosene (Commercial type "A") or a gasoline-kerosene mixture (Commercial type "B"). JP-4, containing 65 percent gasoline, has a somewhat higher flash point, –10°F to 30°F (–23°C to –1°C), and a lower ignition temperature, 468°F (242°C), than gasoline.

Because jet fuels have somewhat lower volatility than gasoline, if they are spilled and ignited the resulting fire may not spread across the surface quite as rapidly as with gasoline. This slowness is particularly true with kerosene-type fuels. However, in a major airplane crash, the impact is sufficient to atomize the spilled liquid. The fuel mist will ignite readily, propagating flames rapidly in spite of any theoretical advantage the lower volatility of kerosene-type fuels might initially present.

Because water will cool kerosene below its flash point, the use of water fog is recommended. Ordinary foam, carbon dioxide, and dry chemical extinguishers will also extinguish kerosene fires. Table 3–4 includes a summary of the characteristics for kerosene.

Naphtha

To a perplexed firefighter lost in a jungle of chemical terminology, there seems to be some truth to the story about a convention of chemists sitting around inventing names for chemicals. "Gentlemen," said one scientist, "here we have fifty completely different, new hydrocarbons. Let's call 'em naphtha."

Naphtha seems to be everyone's favorite name for a liquid. There are naphthas with high flash points and low flash points. There are heavy naphthas and light naphthas. There are naphthas made from petroleum and completely different naphthas made from coal tar. "Stoddard solvent" and "safety solvent" are both called high-flash naphtha or petroleum spirits, or thinner, or mineral turpentine. Lighter fluid is sometimes called petroleum naphtha when it is not being called petroleum ether (which also has a couple of additional names). Varnish makers and painters use a lot of naphtha. There are at least three kinds of "painter's naphtha," each with a different set of characteristics and several names. Therefore, sorting out naphthas according to hazard is difficult.

Petroleum Ether

The name of this liquid also causes trouble. Chemically, petroleum ether is not really an ether, so many chemists object to calling it one. Petroleum ether is often called benzin, but there are objections to this name too. It can be confused with benzene, a different hydrocarbon, so the use of the word *benzin* is actively discouraged although there is an official grade of this liquid called petroleum benzin. **Ligroin** is the correct name for petroleum ether. Hardly anyone uses it, except in the laboratory. As we might guess, ligroin is sometimes called petroleum naphtha.

Whatever you call it, petroleum ether is a clear, colorless, highly flammable liquid with a strong odor. Depending on its manufacture, its flash point ranges upward from –40°F (–40°C) to around 0°F (–17°C). Like all hydrocarbons, it is insoluble in water. Petroleum ether is the lightest liquid fraction of petroleum. Made from pentanes and hexanes, it is used as a solvent in the laboratory; as a solvent in drugs, oils, fats, and waxes; as a fuel; in paints and varnishes; as an insecticide; in photography; and as a detergent. Its use as a spot remover and as a dry cleaner has been discouraged because of its flammability. Ligroin is often shipped and stored in one-gallon (four-liter) glass containers.

Inhaling petroleum ether vapors may cause headache, nausea, and intoxication, with consequent loss of judgment. Firefighters exposed to petroleum ethers should wear self-contained breathing apparatus and protective clothing. Extinguishment of petroleum ether fires with water is a questionable procedure. Use dry chemical or ordinary foam.

Fuel Oils

Many different kinds of fuel oils are in use today. Fuel Oil No. 1 is a synonym for kerosene or range oil; No. 1D is a light diesel fuel. Fuel Oil No. 2 is used as a domestic fuel oil; No. 2D is a medium diesel fuel; No. 3 is no longer a current standard. Numbers 4, 5, and 6 are light, medium, and heavy industrial fuels, respectively. These oils become increasingly viscous from No. 1 thorough No. 6, and different fuel oil burners are designed for oils of a particular viscosity.

Viscosity is the resistance offered by a liquid to flow. As the length of a hydrocarbon chain increases, the viscosity of the resulting liquid also increases. Viscosity assumes firefighting significance in the case of fuel oils. A specialized firefighting method can be used against heavy grade fuel oils, asphalt, and lubricating oils. If oil and water are shaken together, tiny droplets will form on the surface of the liquid. Although the water and the oil have not mixed, water is trapped inside these bubbles. This is called an emulsion. Directing a coarse water spray onto the surface of a viscous liquid whose vapors are burning forms an emulsion that can smother the fire. This emulsion, while temporary in light fuel oils, is very persistent when formed with heavy grade oils. It will even guard against the possibility of a flashback (reignition).

The emulsification technique must be used with care in fires involving high–flash point liquids. Injecting water into fires involving flammable liquids having flash points above 212°F (100°C), the boiling point of water, increases the risk of water's becoming trapped underneath the liquid and being converted to steam. Water, having an expansion rate of 1,700 to 1, significantly increases the possibility of a boilover, spreading the fire and causing severe burns to those in close proximity to the container. Some fire scientists advocate the circulation of cooler oil from the bottom of the tank, lowering the temperature of the surface liquid below its flash point, as a possible method of extinguishment. Dry chemical and foam application have also been used. Most of these comments on fuel oils apply to lubricating oils and asphalt as well.

RING HYDROCARBONS

Besides chains and branched chains, other types of hydrocarbon series are possible. The ever-versatile carbon atom is also capable of forming rings. Three examples of ring hydrocarbons are **Benzene, toluene, and xylene.** Ring hydrocarbons are **unsaturated** hydrocarbons as indicated by the double bonds between some carbon atoms. Being hydrocarbons, they are flammable, but they have an added hazard: a rather high level of toxicity (see Table 3–6).

Benzene (Benzol)

Benzene (Figure 3-3), the most toxic of the ring hydrocarbons, is very widely used in industry. It is used in the manufacture of styrene, phenol, and nylon. Aniline and other dyes are made from benzene. It is used in the production of such insecticides as DDT and benzene hexachloride

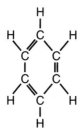

FIGURE 3–3. Benzene (C_6H_6).

TABLE 3–6. Properties of Typical Ring Hydrocarbon Fractions

	Flash Point	Ignition Temp.	Lower Flam. Limits	Upper Flam. Limits	Specific Gravity	Vapor Density	Boiling Point
Benzene (benzol)	12°F −11°C	1044°F 562°C	1.3	7.1	0.9	2.8	176°F 80°C
Naphtha (coal tar)[a]	100°–110°F 38°–44°C	531°F 277°C			0.8		300°–400°F 149°–204°C
Naphthalene (white tar)	174°F 79°C	979°F 526°C	0.9	5.9	1.1	4.4	424°F 218°C
Toulene (toluol)	40°F 4°C	997°F 536°C	1.2	7.1	0.9	3.1	231°F 111°C
Xylene ortho	63°F 17°C	979°F 526°C	1.0	6.0	0.9	3.7	292°F 144°C
meta	77°F 25°C	982°F 528°C	1.1	7.0	0.9	3.7	282°F 139°C
para	77°F 25°C	984°F 529°C	1.1	7.0	0.9	3.7	281°F 138°C

[a]Coal tar naphtha is composed mainly of toluene and xylene.

(Lindane); to make many organic compounds; in the manufacture of medicinal chemicals, artificial leathers, and oilcloth; and as a solvent in varnishes, lacquers, resins, waxes, oils, and such commodities as airplane dope (used in model airplane building). Large amounts of benzene are used in motor fuels, particularly aviation and 100+ octane gasolines.

Benzene can damage bone marrow, producer of the body's blood cells, causing anemia. It is also suspected of producing leukemia (a blood cancer) and of affecting the liver and kidneys. Symptoms of poisoning may not appear until the effect on the body has approached fatal proportions. Benzene has a cumulative action, and prolonged inhalation, such as by workers constantly exposed to its fumes, is considered the most dangerous type of exposure. *Brief high exposures, the major hazard to firefighters, can cause anesthesia, leading to death by respiratory failure.*

Benzene vapors are insidious. They have a pleasant smell and give no adequate warning of their toxic effects. Because of their pleasing smell, members of the ring hydrocarbon series are often called the aromatic hydrocarbons. There is no appreciable irritation or discomfort when benzene is inhaled, even at dangerous levels, but the liquid can be absorbed through the skin in harmful amounts, causing irritation and even blisters. Symptoms of benzene poisoning include dizziness, a tightening of the leg muscles, forehead pressure, intoxication, and eventual coma. Firefighters should protect themselves from aromatic hydrocarbon vapors with self-contained breathing apparatus.

Benzene has an extremely high concentration of carbon in its molecule, as compared with other hydrocarbons, caused by the presence of three double bonds between carbon atoms, which reduces the number of bonds available for hydrogen. Benzene (and other members of the ring series) burns with an extremely smoky flame that is composed of particles of unburned carbon escaping from the combustion zone. The result is the production of large quantities of carbon monoxide.

Although fires involving this clear, colorless liquid can be fought with water spray, this procedure may not be the best. Dry chemical, carbon dioxide, or ordinary foam may give better results.

Toluene (Toluol)

Toluene is simple benzene with an added carbon side chain, as shown in Figure 3–4. This side chain is known as a **methyl group**, and toluene is therefore sometimes called *methylbenzene*. Along with benzene, toluene is generally produced from coal. Toluene is a flammable, colorless liquid with an easily detected benzene odor. It is used in the manufacture of organic compounds; lacquers, aviation gasoline, medicines, dyes, perfumes, and saccharin. As a solvent, it will dissolve gums, resins, some oils, cellulose acetate, and cellulose ether. Trinitrotoluene (TNT) is a nitrated form of toluene. Toluene is shipped in containers ranging from glass bottles to large tanks.

Although it is an eye and respiratory irritant, and although extreme inhalation of vapors may cause paralysis of the respiratory center, acute poisonings from toluene are rare. It is less toxic than benzene but can cause fatigue, weakness, and loss of judgement.

Toluene fires can be fought with water spray although, once again, extinguishment is questionable. Better results may be obtained from the use of dry chemical, CO_2, or ordinary foam.

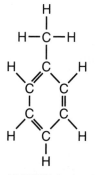

FIGURE 3–4. Toluene (C_7H_8).

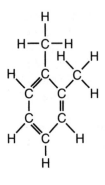 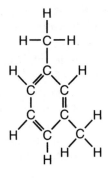 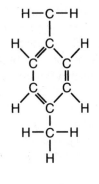

Ortho-xylene	Meta-xylene	Para-xylene
C_8H_{10}	C_8H_{10}	C_8H_{10}

FIGURE 3–5. Isomers of Xylene.

Xylene (Xylol)

Like toluene, xylene is simple benzene except it has two side chains which can be attached at different places as shown in Figure 3–5.

All of the xylenes are clear, colorless, flammable liquids. Their most common use is in aviation gasoline, but xylenes will dissolve lacquers, enamels, and rubber cements. Other uses include the manufacture of dyes, vitamins, medicines, and polyester plastics. Xylenes are less toxic than benzene and give warning of their presence by irritating the eyes, nose, and throat. Vapors in high concentrations are anesthetic, and a few cases of fatal poisoning have occurred. Like the other ring hydrocarbons, xylene burns with an extremely smoky flame, producing carbon monoxide in high concentrations. Firefighting methods parallel those for benzene and toluene.

Turpentine

Turpentine is a major liquid hydrocarbon not derived from either coal or oil. It is not a ring hydrocarbon. Crude turpentine is a sticky yellowish gum obtained from pine trees. From this, balsam oil of turpentine is distilled. Oil or spirits of turpentine is a colorless liquid with a characteristic odor. It is used as a solvent for oils, resins, varnishes, paints, lacquers, rubbers, polishes, and waxes.

Excessive exposure to turpentine fumes can cause nausea and headache. Kidney injury is possible. Its fumes are irritating to the nose and throat. Firefighters should be protected by self-contained breathing apparatus, as burning turpentine produces carbon monoxide.

Oil of turpentine, unlike most hydrocarbons, heats spontaneously (Table 3-7). It reacts vigorously with chlorine and iodine. Its fire characteristics are similar to those of kerosene.

TABLE 3-7. Oil of Turpentine

| Flash Point | Ignition Temp. | Flammable | | Specific Gravity | Vapor Density | Boiling Point | Water Soluble |
		Lower Limits	Upper Limits				
95°F 35°C	488°F 253°C	0.8	—	1.0	4.8	300°F 149°C	No

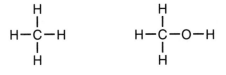

Methane
(CH$_4$)

Methyl alcohol
(CH$_3$OH)

FIGURE 3-6. Methyl alcohol.

ALCOHOL

When discussing combustion, we mentioned that molecule fragments in a combustion zone were called **radicals**. A particular group of atoms inside a molecule that tends to operate as a unit is also called a radical. Notice the 0–H combination on the right side of the structural formula of **methyl alcohol** (Figure 3-6). This O–H combination is known as the **hydroxide** (or hydroxyl) **radical**. When a hydrocarbon adds a hydroxyl radical, it generally changes into an alcohol. Indeed, the combination C–0–H means "alcohol" to a chemist.

When an O–H group is added to an inorganic element or group of elements that does not include a carbon atom, the resulting substance is called a **base.** NaOH, called sodium hydroxide or lye, is a very caustic basic (alkaline) material. NH$_4$OH is ammonium hydroxide. A weak solution of NH$_4$OH is called ammonia water. These compounds *are not* alcohols.

Stripped of the hydroxyl radical, the left side of the methyl alcohol molecule is as shown in Figure 3–7.

This group of four atoms is known as the **methyl radical**. We saw that when benzene added this group of atoms to its molecule, it changed into

FIGURE 3-7. The methyl radical.

toluene. Whenever this combination appears as part of a larger molecule, it is generally reflected in the title of the compound: methyl alcohol, methyl acetate, methyl ether, and so forth.

When other members of the paraffin series are stripped of one of their hydrogen atoms, they also form radicals. When they add a hydroxyl radical, they change into alcohols (see Table 3–8).

Correct chemical terminology reserves the use of the suffix -ol for alcohols. This is why the names *benzol, toluol,* and *xylol* are technically incorrect, although they are frequently used. Benzene, toluene, and xylene are hydrocarbons, not alcohols.

As a group, alcohols generally have lower heats of combustion than equivalent hydrocarbons. Alcohols are partially oxidized, partially burned already. But, as their carbon chain grows longer, so does their heat of combustion. Amyl alcohol burns hotter than methyl alcohol. Both burn hotter than ordinary combustibles. A lengthening carbon chain changes other properties of alcohols. Methyl, ethyl, and the propyl alcohols are completely soluble in water. (The hydroxyl radical, O–H, is very similar to the structure of water, H–O–H.) Beginning with **butyl alcohol**, however, the water solubility of alcohols grows less and less as the hydrocarbon radical assumes greater weight. *Heavier alcohols may be totally insoluble in water.* Eventually, as the hydrocarbon chain grows longer, alcohols become heavier and heavier liquids until, at last, solid alcohols are formed. Cetyl alcohol, with a sixteen-carbon chain, is a white crystalline powder first discovered in a whale.

There are three major types of alcohols. The addition of a single hydroxyl group forms the alcohols shown in Table 3–8. Two hydroxyl groups attached to a chain form glycols. Three hydroxyl groups from glycerol (see Figure 3–8).

TABLE 3–8. Names of Hydrocarbons, Radicals, and Alcohols

Hydrocarbon	Radical	Alcohol	Official Name
Methane	Methyl	Methyl alcohol	Methanol
Ethane	Ethyl	Ethyl alcohol	Ethanol
Propane	Propyl	Propyl alcohol	Propanol
Isopropane	Isprophy	Isopropyl alcohol	Isopropanol
Butane	Butyl	Butyl alcohol	Butanol
Isobutane	Isobutyl	Isobutyl alcohol	Isobutanol
Pentane	Amyl[a]	Amyl alcohol	Pentanol
Hexane	Hexyl	Hexyl alcohol	Hexanol
Heptane	Heptyl	Heptyl alcohol	Heptanol
Octane	Octyl	Octyl alcohol	Octanol
Nonane	Nonyl	Nonyl alcohol	Nonanol
Decane	Decyl	Decyl alcohol	Decanol

[a] The five-carbon radical more often called amyl than pentyl even though the official name for amyl alcohol is pentanol. You might also expect that benzene would form the benzyl radical. It doesn't. Toluene does. Benzene forms the phenyl radical.

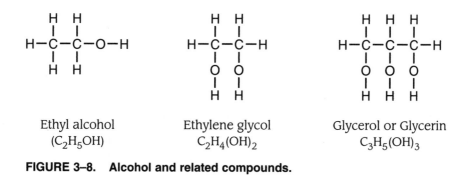

Ethyl alcohol
(C_2H_5OH)

Ethylene glycol
$C_2H_4(OH)_2$

Glycerol or Glycerin
$C_3H_5(OH)_3$

FIGURE 3–8.   Alcohol and related compounds.

Methyl Alcohol (Methanol, Wood Alcohol, Wood Spirits)

More than 150 million gallons of methyl alcohol are produced annually in the United States. In terms of widespread use, it is one of the "big three" alcohols, along with ethyl alcohol and isopropyl alcohol. Were it not for its toxicity, methyl alcohol would be used even more.

Although it was once prepared by heating wood in the absence of air (hence its name "wood alcohol"), 90 percent of all methyl alcohol is now made by the partial oxidation of methane or by combining carbon monoxide and two hydrogen molecules, under heat and pressure, with the aid of catalysts, zinc oxide and chromic acid. (A **catalyst** is a material that speeds or assists a chemical reaction but takes no part in the reaction. Catalysts are frequently employed in industrial processes.) Methyl alcohol is prepared as follows:

$$C=O \quad + \quad \begin{matrix} H-H \\ H-H \end{matrix} \quad \xrightarrow{\text{heat, pressure, catalyst}} \quad H-\overset{\displaystyle H}{\underset{\displaystyle H}{\overset{|}{\underset{|}{C}}}}-O-H$$

Carbon monoxide   **Hydrogen**   **Methyl alcohol**

Methyl alcohol is a clear, colorless, volatile, flammable liquid. When pure, it has a slight alcohol odor. When it is impure, the odor may become stronger. It burns with the typical blue flame of alcohols. Table 3–9 lists the fire characteristics of methyl alcohol. Because it is water soluble, dilution techniques can be used for extinguishment. In addition, water fog, in conjunction with dry chemical, "alcohol" foam, and carbon dioxide have all been used successfully.

Deaths have occurred from drinking small amounts of methyl alcohol. Inside the body, it is oxidized into formaldehyde and formic acid, both which are poisonous, and it is eliminated slowly. Inhalation of methyl alcohol vapors is also hazardous. Although single exposures to fumes in low concentrations may cause no harmful effects, heavier concentrations can be dangerous. Depending on the amount of exposure, poisoning may

TABLE 3–9. Properties of Various Alcohols, Glycols, and Gycerol

	Flash Point	Ignition Temp.	Flammable Limits (percent by volume in air)		Specific Gravity (water=1.0)	Vapor Density (air=1.0)	Boiling Point	Water Soluble
			Lower	Upper				
Allyl alcohol[a] (Propenyl alcohol)	70°F 21°C	713°F 378°C	2.5	18.0	0.9	2.0	206°F 97°C	Yes
U.L. Class 40: Amyl alcohol (Pentanol)[b] (normal)	91°F 33°C	572°F 300°C	1.2	10.0	0.8	3.0	280°F 138°C	Slightly
U.L. Class 40: Butyl alcohol (Butanol) (normal)	84°F 29°C	650°F 343°C	1.4	11.2	0.8	2.6	243°F 117°C	Yes
(secondary)	75°F 24°C	763°F 406°C	1.7	9.8	0.8	2.6	201°F 94°C	Yes
(tertiary)	52°F 11°C	892°F 478°C	2.4	8.0	0.8	2.6	181°F 83°C	Yes
U.L. Class 70: Denatured alcohol[c]	60°F 16°C	750°F 399°C			0.8	1.6	175°F 79°C	Yes
U.L. Class 70: Ethyl alcohol[d] (Ethanol, Grain alcohol)	55°F 13°C	793°F 423°C	4.3	19.0	0.8	1.6	173°F 78°C	Yes
Ethylene glycol (Glycol)	232°F 111°C	775°F 413°C	3.2		1.1		387°F 197°C	Yes
U.L. Class 10–20: Glycerol (Glycerin)	320°F 160°C	739°F 393°C			1.3		554°F 290°C	Yes

continued

46

TABLE 3–9. Properties of Various Alcohols, Glycols, and Gycerol (cont.)

	Flash Point	Ignition Temp.	Flammable Limits (Percent by volume in air)		Specific Gravity (Water=1.0)	Vapor Density (Air=1.0)	Boiling Point	Water Soluble
			Lower	Upper				
U.L. Class 35–40: Isoamyl alcohol[e] (Fuel oil)	109°F 43°C	657°F 347°C	1.2	9.0 (at 212°F or 100°C)	0.8	3.0	270°F 132°C	Slightly
U.L. Class 40–45: Isobutyl alcohol (Isopropyl carbinol)	82°F 28°C	800°F 427°C	1.7 (at 212°F or	10.9 (at 100°C)	0.8	2.6	225°F 107°C	Yes
U.L. Class 70: Isopropyl alcohol (Isopropanol)	53°F 12°C	750°F 399°C	2.1	13.5	0.8	2.1	181°F 83°C	Yes
U.L. Class 70: Methyl alcohol (Methanol, Wood alcohol)	52°F 11°C	867°F 464°C	7.3	36.0	0.8	1.1	147°F 64°C	Yes
U.L. Class 55–60: Propyl alcohol[d] (Propanol)	77°F 25°C	700°F 371°C	2.1	13.5	0.8	2.1	207°F 97°C	Yes
Propylene glycol (Methyl ethylene glycol)	210°F 99°C	790°F 421°C	2.6	12.5	1.0 (or over)		370°F 188°C	Yes

[a]Allyl alcohol has a pungent odor somewhat like that of mustard. It is poisonous and very dangerous to the eyes. Absorbed readily through the skin, highly toxic by inhalation or ingestion. Odor may not be sufficient to warn fire fighters to don protective equipment! Used in several industrial processes and as a military poison gas. Shipped under a poison "B" label.

[b]An isomer of amyl alcohol, with a lower boiling point of 245°F (118°C) has a classification of 40—45. This isomer has a flash point of 94°F (34°C).

[c]Different government formulations can lower the flash point to less than 20°F (–7°C).

[d]Water mixtures have higher flash points.

[e]There are two other isomers of isoamyl alcohol, one of which has a flash point of 67°F (19°C).

47

cause a variety of symptoms: headache, fatigue, nausea, acidosis, dizziness, intoxication, unconsciousness, a sighing respiration, respiratory failure, and circulatory collapse.

The greatest danger from methyl alcohol is its specific toxic effect upon the eyes: It can atrophy the optic nerve and retina. Visual symptoms, such as blurring of vision, clear up, only to recur and progress to blindness. Exposure to methyl alcohol liquid may cause the skin to become dry and cracked, the poisoning can occur through breaks in the skin.

Flammable liquids are mainly used as fuels and solvents. Methyl alcohol is a better solvent than ethyl alcohol. It is abundant and cheap, and it has a low boiling point, so it can be easily vaporized after it is no longer needed. As a consequence, in spite of its toxicity (it cannot be made nontoxic), methyl alcohol is widely used in industry. It is a solvent for surface coatings of various kinds and for paint removers, inks, and adhesives. It is used in the production of formaldehyde and the synthesis of various chemicals. It is sometimes used as an antifreeze in gasolines and diesel oils, where it combines with the water in these fuels to prevent the formation of ice. Other uses include manufacture of medicinals, softener for pyroxylin plastics, and as a fuel in such appliances as picnic stoves. Methyl alcohol is shipped in a wide variety of containers.

Ethyl Alcohol (Ethanol, Grain Alcohol, Alcohol, Cologne Spirits)

Ethyl alcohol, mixed with water and flavorings, seems to be everybody's favorite flammable liquid. It is by far the most important industrial alcohol, so much so that the amount of ethyl alcohol sold is a direct reflection of the U.S. economy. It is used to manufacture a variety of industrial compounds, such as aldehydes, acids, ethers, acetates, drugs and medicinals (tincture of iodine is iodine dissolved in ethyl alcohol), dyes, certain rubbers, and explosives. It is used in food flavorings and extracts, cosmetics, and lotions of various kinds. It is a solvent for resins and varnishes. Small wonder the general term *alcohol* has come to mean ethyl alcohol, although there are many different alcohols.

Sugars found in such substances as grain, fruits, and molasses, when added to water and yeast, will ferment into "grain alcohol," while giving off carbon dioxide.

$$C_{12}H_{22}O_{11} + H_2O \xrightarrow{\text{yeast}} 4\ O_2H_5OH + 4\ CO_2$$

Sugar Water   **Ethyl   Carbon**
                                                                alcohol dioxide

Ethylene, a flammable gas, is used in the two major industrial processes for the production of synthetic ethyl alcohol. The most important method adds a water molecule to ethylene through the use of heat, pressure, and a catalyst:

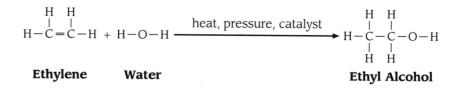

Ethylene   **Water**   **Ethyl Alcohol**

Industrial processes produce pure ethyl alcohol, sometimes called "absolute alcohol" or **anhydrous (water-free) alcohol,** a colorless, volatile, flammable liquid with a familiar odor (some are odorless) and a burning, pungent taste. Absolute alcohol is so water soluble that it absorbs water from the air. Because ethyl alcohol is totally water soluble, firefighting methods like dilution can be very effective. As additional water is added to ethyl alcohol (about six parts to one), the flash point will rise above 100°F (38°C). This property is useful in diluting or extinguishing a spill. Adding sufficient water creates a mixture that is difficult to ignite (see Table 3–10). Dilution techniques are less effective on tanks because of the tank's limited capacity.

The inhalation of ethyl alcohol vapors is anesthetic, but is not particularly toxic. Some eye irritation can occur at higher concentrations. The ingestion of ethyl alcohol can cause nausea, vomiting, flushing, drowsiness, toxication, even coma or stupor. Drinking ethyl alcohol in large amounts can cause respiratory collapse and death. However, unlike methyl

TABLE 3–10.   Dilution Effect on Flash Point of Ethyl Alcohol

Percentage of Alcohol	Proof	Flash Point	
		Fahrenheit	*Centigrade*
100%	200	55°	13°
96	192	62°	17°
95	190	63°	17°
80	160	68°	20°
70	140	70°	21°
60	120	72°	22°
50	100	75°	24°
40	80	79°	26°
30	60	85°	29°
20	40	97°	36°
10	20	120°	49°
5	10	144°	62°

alcohol, ethyl alcohol is oxidized fairly rapidly by the body into nontoxic products, carbon dioxide, and water. There is no cumulative effect.

When ethyl alcohol is not used as a beverage or in cosmetics, various denaturants are added to it. Denaturants can include methyl alcohol, camphor, amyl alcohol, gasoline, isopropyl alcohol, turpentine, benzene, castor oil, acetone, nicotine, aniline dyes, ethers, sulfuric acid, kerosene, and other liquids not often recommended for drinking. These denaturants are generally chosen because they have boiling points similar to that of ethyl alcohol and cannot be easily separated by distilling. They can make pure ethyl alcohol more toxic and flammable.

Other Alcohols

To complete the discussion of various alcohols, a closer look at **iso-propyl alcohol,** rubbing alcohol, which also has important industrial uses, is necessary. It is not considered toxic, although it can be locally irritating, and, in high concentrations, act as a narcotic.

Ethylene glycol, widely used as a permanent antifreeze and motor coolant, is not considered highly toxic either, unless someone decides to drink it. A mixture of ethylene glycol and water will freeze at a lower temperature than either liquid separately.

Glycerin (glycerol) is often used in the making of candy, and the only health hazards it presents seem to be those associated with overeating.

Table 3–9 shows the fire characteristics of these three liquids.

The following incident involving isopropyl alcohol is an example of the problems faced when combating fires involving alcohols.

On September 24, 1990 in Tampa, Florida, the fire department received a 911 call for a fire in a chemical plant. Due to the large volume of dense black smoke, first-arriving units questioned whether alcohol was burning or not (a burning alcohol flame, by itself, is almost colorless). However, after checking with plant officials it was confirmed that the material involved was isopropyl alcohol. The fire started as the result of an explosion and rupture of a 55,000 gallon isopropyl alcohol tank, killing one worker. The tank lid traveled approximately 50 yards from the tank. The fire was burning at temperatures in excess of 1,500 degrees Fahrenheit. Two adjoining tanks were involved in the fire with direct flame impingement on both. Eventually, fire department personnel were told that both tanks were empty and had been purged of any hazardous contents. However, as a precaution, they continued to consider them to be full or empty with flammable vapors present. The initial attack was to protect exposures while setting up the ability to place into operation two ground monitors to apply alcohol foam at a rate of 750 gallons per minute (regular flammable liquid foam will break down when it comes in contact with alcohols). Shortly after the operations began one of the engines supplying a significant portion of the water supply broke a pump shaft. Some delays were experienced until the engine could be replaced and the hose lines put back into operation. Once the foam operation was re-initiated, confinement was experienced in approximately three hours.

ETHERS

The smell of ether is familiar to anyone who has been in a hospital, and we are aware of the anesthetic properties of most ethers. Their vapors can produce unconsciousness (when such a state is desirable) or even death from respiratory failure, if the exposure is severe.

Dioxane, diethylene ether, has an irritating and poisonous vapor. Toxic amounts of liquid dioxane can be absorbed through the skin. Besides their use as anesthetics, ethers have a wide industrial value. Table 3–11 lists some of them.

As a class of flammable liquids, ethers present several specialized fire problems. Note the low boiling points of many. **Dimethyl ether** (see Table 3–11), which boils at –11°F (24°C), is a flammable gas and is included here only for comparison. **Methyl ethyl ether** and **ethyl ether** boil at such low temperatures that the heat of a human hand can turn them into a gas. (This is *not* a recommended experiment). As a consequence, fires involving these two ethers may be extremely difficult to extinguish. The amount of vapor they produce will be impressive. Their **ebullition,** or bubbling when heated, may cause the breakdown of any foam blanket placed over them. Most ethers have a specific gravity less than that of water and will float. Like alcohols, some are soluble in water, some are not; this requires identification of the particular ether involved.

The low ignition temperatures of many ethers means that the slightest ignition source can result in an explosion of the vapors. This is why static electricity is such a hazard in hospital operating rooms and why potential ignition sources must be rigidly controlled. (This subject will be discussed more fully when flammable anesthetics are investigated in Chapter 5.)

Ether Peroxides

Over the past several years, the danger of **ether peroxide** formation has received a great deal of publicity. There have been several incidents in which aging ether was discovered (particularly in colleges), taken to a supposedly safe place, and detonated. The resulting explosions were far greater than the initiating charge. On a few occasions, suspected bottles were taken to a vacant lot where stones were thrown at them. When struck, many of the bottles exploded. A doctor was killed when an ether bottle he was trying to open exploded and disemboweled him. Peroxide formation, the addition of oxygen atoms to the ether molecule, can be accelerated by several factors: heat, distillation, extended storage, and exposure to light or air. For this reason most ethers are shipped in amber bottles or cans. Contact with certain metals, particularly iron and copper, appears to inhibit this formation, and so a piece of iron wire is often put inside ether cans. However, even in closed cans, formation of peroxides

TABLE 3–11. Uses and Properties of Ethers

	Flash Point	Ignition Temp.	Lower Flammable Limit (% by vol. in Air)	Upper Flammable Limit (% by vol. in Air)	Specific Gravity (Water=1.0)	Vapor Density (Air=1.0)	Boiling Point	Water Soluble	Comments
Butyl ether (Dibutyl ether)	77°F 25°C	382°F 195°C	1.5	7.6	0.8	4.5	286°F 141°C	No	Used as an industrial solvent. Has a mild odor. Avoid breathing vapor or skin contact with liquid.
Dimethyl ether (Methyl ether)	Gas	662°F 350°C	3.4	18.0		1.6	-11°F -24°C	Yes	Often shipped as liquid in pressurized cylinders. Uses: spray propellant, refrigerant, solvent. Requires DOT red label: fl. gas.
Dioxane—1, 4 (Diethylene oxide)	54°F 12°C	356°F 180°C	2.0	22.0	1.0	3.0	214°F 101°C	Yes	Irritating and poisonous vapor. Toxic amounts of liquid may be absorbed through the skin. Uses: industrial solvent, paint remover.
Ethyl ether (Diethyl ether; Diethyl oxide; Ethyl oxide; Ether)	-49°F -45°C	356°F 180°C	1.9	48.0	0.7	2.6	95°F 35°C	Slightly	Not corrosive or poisonous, but an anesthetic. Used as anesthetic, in manufacture of smokeless powder, as a solvent, in motor fuels, etc. Very low boiling point. Has a UL classification of 100.
Isopropyl ether (Diisopropyl ether)	-18°F -28°C	830°F 443°C	1.4	21.0	0.7	3.5	156°F 69°C	Slightly	Similar to ethyl ether in toxicity. Forms peroxides readily. Uses: solvent for oils, waxes, resins, dyes, to make rubber cements.
Methyl ethyl ether	-35°F -37°C	374°F 190°C	2.0	10.1	0.7	2.1	51°F 11°C	Yes	Note the boiling point and ignition temperature.
Vinyl ether (Divinyl ether)	-22°F -30°C	680°F 360°C	1.7	27.0	0.8	2.4	102°F 39°C	No	Wide anesthetic use. Can form peroxides. Low boiling point.

is possible. This is particularly true for isopropyl ether or absolute (pure) ether, although all ethers are suspect. Note that organic peroxides (ether peroxide is one of these) are generally classed as very unstable chemical compounds. (See Chapter 11 for further discussion.)

Storage and Shipping Conditions

The California State Fire Marshal's Office distributes the following excellent summary of safety measures for those who use ether.

1. Glass containers of all sizes should be avoided whenever possible.
2. All containers should be dated so that the age of the contents may be determined.
3. Isopropyl and absolute ethers should not be kept for more than six months; ethyl and other ethers for not more than one year.
4. Ether should be stored in as cool a location as feasible (but not stored in refrigerators unless explosion-proof).
5. Ether should always be tested for peroxide content before any distillation and, of course, should not be used if peroxides are found to be present.
6. Do not attempt to open any container of uncertain age or condition, or whose cap or stopper is tightly stuck.
7. Manufacturers should be contacted to learn any general recommendations regarding safe handling in storage and use, and any specific recommendations for the addition of inhibitors to prevent peroxide formation. Manufacturers can also recommend the best methods of chemical testing to detect peroxide content and methods for removing peroxides by chemical means.

Special precautions are necessary when ether is used as an anesthetic. As a matter of practice, ether is not used for this purpose if a can has been open longer than twenty-four hours. All of this brings to mind a disquieting picture of countless bottles and cans of ether, slowly turning dangerous on dusty rear shelves, in hospitals and schools around the country. Firefighters called upon to dispose of a suspected container of ether should obtain technical help, although there have been no reported cases of spontaneous explosion upon gentle handling.

Remember: Ether peroxides may not be visible as crystals. Treat any elderly ether as potentially explosive! After a fire, any bottles or cans exposed to heat should be discarded.

Finally, the name *ether* signifies to the chemist that the following lineup of atoms occurs in the molecule: carbon—oxygen—carbon. The various names given them—**diethyl** (two ethyls) ether or **dimethyl** (two methyls) ether—merely show how the entire molecule looks. (See Figure 3-9.)

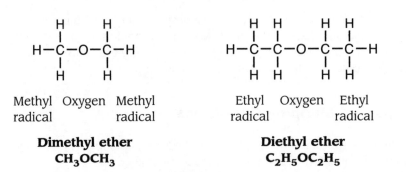

Methyl Oxygen Methyl
radical radical

Dimethyl ether
CH₃OCH₃

Ethyl Oxygen Ethyl
radical radical

Diethyl ether
C₂H₅OC₂H₅

FIGURE 3–9. Ether molecules. [*Dimethyl ether and ethyl alcohol have exactly the same number of atoms of the same kind, two carbons, six hydrogens, and a single oxygen (C_2H_6O). The difference in their properties is entirely due to the interior arrangement of the atoms.]

AMINES

Several other important classes of flammable liquids remain. One of these, the amines, is a large group of colorless to yellow flammable liquids and gases, derived from ammonia, NH_3. When one or more of the hydrogen atoms in ammonia is replaced by a carbon-containing radical, an amine results (see Figure 3–10).

Most of the amines retain the aroma of ammonia, although a few smell more like rotten fish. Modern industry puts them to seemingly endless uses: in the manufacture of specialty soaps, dyestuffs, rubber chemicals, textile specialties, corrosion inhibitors, insecticides, drugs, wash-and-wear resins, fungicides, herbicides, petroleum additives, rubless floor polishes, shampoos, and so on.

As Table 3–12 shows, several of the amines present extraordinary hazards. **Ethyl amine** and **isopropyl amine** have boiling points so low that they turn to gases at the slightest opportunity. Great care must be exercised when drums are opened. Refrigerated storage may be necessary. **Hydroxyl amine** explodes at 265°F (129°C). Remote fire-fighting methods and unstaffed large appliances should be employed and the surrounding area evacuated.

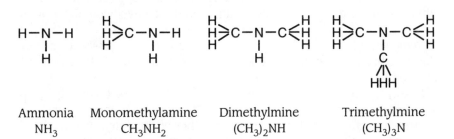

Ammonia   Monomethylamine   Dimethylmine   Trimethylmine
NH_3 CH_3NH_2 $(CH_3)_2NH$ $(CH_3)_3N$

FIGURE 3–10. Amine formulas.

TABLE 3–12. Amines

	Flash Point	Ignition Temp.	Lower Flammable Limit (% by vol. in air)	Upper Flammable Limit (% by vol. in air)	Specific Gravity (water=1.0)	Vapor Density (air=1.0)	Boiling Point	Comments
Allyl amine	-20°F -29°C	705°F 374°C	2.2	22.0	0.8	2.0	128°F 53°C	Irritating and poisonous. Attacks rubber. Dangerous fire hazard.
Butyl amine	45°F 7°C	594°F 312°C	1.7	9.8	0.8	2.5	172°F 78°C	Liquid causes eye injury and skin irritation. Vapor harmful.
Cyclohexyl amine	90°F 32°C	560°F 293°C			0.9	3.4	274°F 134°C	Severe irritant, burns skin on contact. Vapors are nauseating.
Diethylene triamine	215°F 102°C	750°F 399°C			0.95		404°F 207°C	Danger! Liquid causes eye injury and skin burns. Vapor harmful.
Diisopropyl amine	30°F -1°C				0.7	3.5	183°F 84°C	Liquid may cause eye injury and skin irritation. Vapor harmful.
Ethyl amine	0°F -18°C	723°F 384°C	3.5	14.0	0.8	1.6	62°F 17°C	Extremely flammable, with a very low boiling point. Drums should not be opened in temperatures above 60°F (16°C), and opened slowly below that. Breathing of vapor very hazardous. Liquid causes eye injury and skin burns. This material has a severe fire hazard.
Ethylene diamine	110°F 43°C				9.0	2.1	241°F 116°C	Warning! Liquid causes eye injury and skin burns. Breathing of vapor may be harmful. Must carry a DOT poison label: Class B. Individuals exposed to this material may remain sensitive to it.
Hexyl amine	85°F 29°C				0.8	3.5	269°F 132°C	Only slightly water soluble. Liquid causes eye injury and skin burns. Can be absorbed through the skin in harmful amounts.[a]

continued

TABLE 3–12. Amines (continued)

	Flash Point	Ignition Temp.	Lower Flammable Limit (% by vol in Air)	Upper Flammable Limit (% by vol in Air)	Specific Gravity (Water=1.0)	Vapor Density (Air=1.0)	Boiling Point	Comments
Hydroxl amine (oxammonium)	Explodes at 265°F (130°C.)				1.2		158°F 70°C	White crystals or a colorless liquid. Melts at 92°F (33°C). Use caution in approaching. Fight fires remotely. Moderate toxic hazards.
Methyl amine (mono)	Gas	806°F 430°C	4.9	20.7		1.1	21°F -6°C	Flammable gases. Liquid solution (25 percent to 40 percent methylamine in water) also flammable. Eye, skin, and respiratory irritants. Direct or sustained contact causes burns. Dry chemical or CO_2 extinguishers.
(di)	Gas	806°F 430°C	2.8	14.4		1.6	45°F 7°C	
(tri)	Gas	374°F 190°C	2.0	11.6		2.0	38°F 3°C	
Propyl amine (n)	-35°F -37°C	604°F 318°C	2.0	10.4	0.7	2.0	120°F 49°C	Extremely flammable. Breathing of vapor hazardous or fatal. Severe eye, skin, and respiratory irritant.
(iso)	-35°F -37°C	756°F 402°C			0.7	2.0	89°F 32°C	
Triamyl amine	215°F 102°C				0.8		453°F 234°C	Not water soluble. Eye, skin, and respiratory irritant.
Tributyl amine	187°F 86°C				0.8	6.4	417°F 214°C	Not water soluble. Eye, skin, and respiratory irritant.

[a] Skin absorption is also possible with trimethylene tetramine and tetraethylene pentamine and other amines not listed in this table.

Many of the amines are highly toxic. Some of the liquids will burn the skin; some will penetrate it. All of the vapors should be avoided by the use of self-contained breathing apparatus.

Furthermore, people exposed to **ethylene diamine** can build up a sensitivity that renders them vulnerable to its effects. Do not give the material a second chance.

Precautionary labels on amines contain such language as: "Do not breathe vapor. Do not get in eyes, on skin, or on clothing. In case of contact with eyes, skin, or clothing, immediately flush with plenty of water for at least fifteen minutes; for eyes get medical attention. Remove contaminated clothing and shoes at once, and wash thoroughly before re-use."

ALDEHYDES, KETONES, AND ESTERS

The **aldehydes** and the **ketones** are chemically related. They both contain a carbon and an oxygen atom linked by a double bond. This is called a **carbonyl group**. Aldehydes and ketones are called carbonyl compounds.

Surrounding this double bond in the ketones are the various radicals that make each of them different. Acetone (dimethyl [two methyls] ketone) has the structure shown in Figure 3–11.

In addition to acetone, methyl ethyl ketone (propanone; commonly called M.E.K) should be given special attention because of its wide use in industry as a degreasing and defatting agent (see Figure 3–12). This important property also makes it very irritating to the skin.

Other major ketones are widely used in industry as solvents for vinyl resins, esters, ethers, nitrocellulose lacquers, and pharmaceuticals, and to dewax lubricating oils. They assist in the manufacture of resins, dyes, plasticizers, and insecticides. Table 3–13 includes a summary of their fire hazard characteristics. Their flammability is generally of more concern than their toxicity.

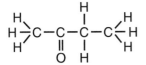

FIGURE 3–11. Acetone (dimethyl ketone) (CH_3COCH_3).

FIGURE 3–12. Methyl ethyl ketone ($CH_3COC_2H_5$).

TABLE 3–13. Properties of Some Aldehydes, Ketones, and Esters

	Flash Point	Ignition Temp.	Lower Flammable Limit (% by vol. in Air)	Upper Flammable Limit (% by vol. in Air)	Specific Gravity (Water=1.0)	Vapor Density (Air=1.0)	Boiling Point	Water Soluble	Comments
Acetaldehyde (Ethanal, Acetic aldehyde)	-36°F -38°C	365°F 185°C	4.1	55.0	0.8	1.5	70°F 21°C	Yes	Colorless liquid, unpleasant fruity odor, irritating and narcotic. Very reactive and potentially unstable.
Benzaldehyde	148°F 64°C	377°F 192°C			1.1	3.7	355°F 179°	No	Water may blanket fire. Low toxicity.
Butyraldehyde (n) (butanol)	20°F -7°C	446°F 230°C	2.5		0.8	2.5	169°F 76°	No	Pungent odor. Low inhalation toxicity. May produce eye and skin burns.
(iso)	-40°F -40°C	490°F 254°C	1.6	10.6	0.8	2.5	142°F 61°C	Slightly	
Formaldehyde (gas) (37% solution)	185°F 85°C	806°F 430°C	7.0	73.0	1.0	1.0	-6°F -21°C	Yes	Eye, skin, and respiratory irritant. Often found in water solutions.
	122°F 50°C						214°F 101°C		
Paraldehyde	96°F 36°C	460°F 238°C	1.3		0.99	4.5	255°F 124°C	Slightly	Freezes at 54°F (12°C). Sedative with hypnotic effects. Poisonings rare.
Propionaldehyde (propanal)	15°F -9°C		3.7	16.1	0.8	2.0	120°F 49°C	Slightly	Unpleasant fruity odor. Toxic when inhaled; avoid breathing vapors.
Acetone (Dimethyl ketone)	0°F -18°C	1000°F 538°C	2.6	12.8	0.8	2.0	134°F 57°C	Yes	Low toxicity. Familiar minty odor. UL Classification of 90.
Mesityl oxide	87°F 31°C	652°F 344°C			0.9	3.4	266°F 130°C	Slightly	Toxic when inhaled or by skin contact.

continued

TABLE 3–13. Properties of Some Aldehydes, Ketones, and Esters (continued)

	Flash Point	Ignition Temp.	Lower Flammable Limit (% by vol. in Air)	Upper Flammable Limit (% by vol. in Air)	Specific Gravity (Water=1.0)	Vapor Density (Air=1.0)	Boiling Point	Water Soluble	Comments
Methyl ethyl ketone (M.E.K.)	21°F -6°C	960°F 516°C	1.8	10.0	0.8	2.5	176°F 80°C	Yes	Smells like acetone. UL rating: 85-90. Somewhat irritating to eyes.
Methyl isobutyl ketone	73°F 23°C	860°F 460°C	1.4	7.5	0.8	3.4	250°F 121°	Slightly	Vapors are irritating and may become narcotic at high concentrations.
Methyl propyl ketone	45°F 7°C	941°F 505°C	1.5	8.0	0.8	3.0	216°F 102°C	Slightly	Vapors are irritating and may become narcotic at high concentrations.
Amyl acetate (iso)	77°F 25°C	713°F 379°C	1.0	7.5	0.9	4.5	290°F 143°C	Slightly	Banana oil. Somewhat irritating.
Butyl acetate (n)	72°F 22°C	790°F 421°C	1.7	7.6	0.9	4.0	260°F 127°C	Slightly	UL Classification of 50-60. Avoid prolonged contact with skin.
(iso)	64°F 18°C	793°F 423°C			0.9	4.0	244°F 118°C	No	
Ethyl acetate (Acetic ether)	24°F -4°C	800°F 427°C	2.5	9.0	0.9	3.0	171°F 77°C	Slightly	UL Classification of 85-90. Sale is controlled by government. Not toxic.
Ethyl silicate	125°F 68°C				0.9	7.2	334°F 168°C		Decomposes in water. Can cause extreme irritation to eyes and nose.
Methyl vinyl acetate	60°F 16°C				0.9	3.5	207°F 97°C	Slightly	Carries a DOT red label when shipped. Low order of toxicity.
Propyl acetate (n)	58°F 14°C	842°F 450°C	2.0	8.0	0.9	3.5	215°F 102°C	Slightly	Somewhat irritating and toxic. High concentrations of vapor to be avoided.
(iso)	40°F 4°C	860°F 460°C	1.8	8.0	0.9	3.5	194°F 90°C	Slightly	

A spectacular fire involving storage tanks holding alcohols, ketones, and esters swept through a San Pedro Harbor loading dock area on August 8, 1972. The fire, described by a Los Angeles City Fire Department deputy chief as being "the worst chemical fire I've ever seen," caused an estimated $500,000 worth of damage. One explosion sent a tank soaring high into the air, trailing flames like a huge rocket. It landed near a warehouse.

The blaze was believed to have been caused by a large tank truck that had entered the area to load up with vinyl acetate—an ester. The truck bumped into a loading dock and caught fire. Flames quickly spread to a nearby 30,000-gallon tank, and the resulting fire eventually involved twenty similar tanks before being brought under control four hours later.

This fire was the first time AFFF (Aqueous Film Forming Foam) was employed on other than an aircraft fire. Its rapid extinguishment by means of AFFF was even more spectacular than the fire.

The aldehydes also contain a carbonyl group. The name *aldehyde* means *al*cohol *dehydr*ogenated because an aldehyde can be produced by subtracting a couple of hydrogen atoms from an alcohol. Notice the difference between ethyl alcohol and acetaldehyde in Figure 3–13.

Aldehydes have an enormous variety of industrial uses, including leather tanning, imparting wet strength to paper, sedatives, embalming, plastics manufacture, and food flavoring. Two aldehydes, acetaldehyde and formaldehyde, are of particular interest.

Acetaldehyde

Acetaldehyde is a colorless liquid with a sickening fruity odor. Its boiling point is extremely low: 69°F (21°C). It will vaporize completely in a normally heated room. Acetaldehyde will react, often violently, with hydrogen cyanide, hydrogen sulfide, anhydrous ammonia, oxidizing agents, alcohols, ketones, caustic alkalis, acid anhydrides, phenols, halogens, amines, and others. Quite a list! It must be isolated from all of these compounds and elements in storage. On top of this, acetaldehyde can react with itself, forming explosive and unstable peroxides or by polymerization (forming long chain molecules). Acetaldehyde, an eye, skin, and respiratory irritant, is capable of producing severe eye burns.

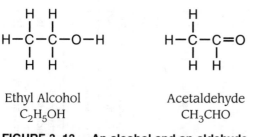

Ethyl Alcohol
C_2H_5OH

Acetaldehyde
CH_3CHO

FIGURE 3–13.   An alcohol and an aldehyde.

There is one bright spot in an otherwise dark picture: Acetaldehyde is totally soluble in water. Flooding dilutes it to a point at which it will not support combustion.

Formaldehyde

Formaldehyde is a colorless gas with an unforgettable, highly irritating odor. We include this gas here because it is often found in a water solution containing from 37 percent to 55 percent formaldehyde, stabilized against self-inflicted reactions by the addition of various percentages of methyl alcohol (usually 15 percent). The flash point of this solution, called formalin, is 122°F (50°C).

Formaldehyde is an eye, skin, and respiratory irritant. The presence of 650 parts per million in the air is an immediate hazard to life. Fires in formalin solutions yield to flooding with water. Table 3–13 summarizes the hazards of several aldehydes.

Esters

The esters are a large group of flammable liquids, often with pleasant odors reminiscent of bananas, pears, pineapples, apples, and other fruit. Esters are used as food flavorings, solvents, in inks, antibiotics, explosives and vitamins—a striking combination! (See Table 3–13).

ANIMAL AND VEGETABLE OILS

Up to this point, we have discussed a sizable number of flammable liquids. Some may be unfamiliar, with the result that one may ask why they are so important. The answer should be clear. The fire potential and toxicity hazards of familiar industrial and household liquids, such as paints, varnishes, lacquers, glues, inks, and paint removers, often depend upon the solvent within them. Most of the liquids discussed in this chapter are used as solvents; hundreds of different solvents are needed to dissolve the multitude of solids that industry wants to use and sell in a liquid state. A discussion of the major constituent of paints, inks, and glues—**linseed oil**—will conclude this section on liquids. Actually, linseed oil is not a solvent but a drying oil. No paint or ink can be complete without something that will cause it to dry and harden upon exposure to air.

Linseed oil has only one real hazard. Since the primary purpose of linseed oil is to oxidize in air, it is subject to spontaneous heating. In Chapter 1, a pile of rags soaked in linseed oil was used as an example of spontaneous ignition. Most flammable liquids are *not* subject to spontaneous heating. Of all those surveyed in this chapter only one has this tendency—turpentine. Petroleum products are not guilty, although often accused. Yet, many fire officers have a mistaken perception about this fire cause. Before marking

TABLE 3–14.   Relative Hazards of Animal and Vegetable Oils

High Hazard	Moderate Hazard	Low Hazard
Cod liver oil	Corn oil	Black mustard oil
	Cotton seed oil	Castor oil
Colors in oil	Olive oil	Coconut oil
	Paints[a]	Lanolin
Fish oil	Pine oil	Lard oil
Linseed oil	Red oil	Oleic acid
	Soybean oil	Oleo oil
Menhaden oil	Tung oil	Palm oil
	Whale oil	Peanut oil
Perilla oil		Turpentine

[a] Paints that contain a dry oil.

down "spontaneous ignition" as the probable cause in a fire report, one should consider whether the suspected liquid came directly from a **living source**. If the fire did not involve one of the liquids listed in Table 3–14, investigate its "cause" further.

The hazards faced when combating fires in buildings where inks and paints are manufactured was never more evident than in the following incident in Cincinnati, Ohio.

At approximately 3:00 P.M., on July 19, 1990, the Cincinnati Fire Department responded to an explosion and fire in the resin yard of the BASF plant. Upon arrival, fire crews were faced with a mass casualty situation as well as a major hazardous materials fire. This plant manufactured coatings and inks using various hazardous materials in the processes. Responding units were encountering victims from the fire three blocks from the initial incident. The plant itself measured two blocks long by one block wide. The main body of the fire was located in the center of the plant. The initial explosion damaged most of the plant and some of the surrounding residential area. After the initial rescue operations were complete, firefighting operations centered around a defensive attack. Numerous explosions were experienced as the fire consumed unknown quantities and types of hazardous materials. Of particular concern to fire department officials were the properties of the newly forming chemicals resulting from the mixing of the chemicals known to exist before the fire. In all, one person died as a result of the incident and seventy-two people were injured.

SUMMARY

The first three chapters of this book have been dedicated to defining the hazards of flammable materials. Flash point, ignition temperature, and a host of other terms relevant to a compound flammability have been discussed thoroughly. However, never let the fact that a material is flammable interfere with gaining an understanding of all the other characteristics of a compound. There are incidents where the flammability of a material is of little consequence with respect to the toxic and poisonous characteristics of the material. Having said that, never forget that the material being dealt with is flammable and will ignite, sometimes with great force, given certain circumstances. One such situation occurred when a fire department responded to an incident at a chemical plant. A worker cleaning an empty tank that had contained toluene was overcome by fumes, rendered unconscious, and fell to the bottom of the tank. Fire department personnel attempted to enter the tank through the access hatch at the top but were unable to fit through the hole while wearing their breathing apparatus. Knowing that toluene was highly toxic, they would not remove their breathing apparatus to enter the tank. A plan was then formulated to cut an access hole into the side of the tank to gain entry and recover the unconscious worker. A metal cutting saw was used with a water spray protection line to effect the rescue. Approximately halfway through the operation an explosive range between the toluene and air was reached and sparks from the cutting operation ignited the vapors. The resulting explosion sent several firefighters to the hospital, causing the death of a veteran captain. The unconscious worker also died, presumably as the result of being overcome by the toxic fumes. Remember, understand all the chemical characteristics of a flammable hazardous material and consider the possible consequences of actions when formulating strategies and tactics.

REVIEW QUESTIONS

1. Gasoline has a flash point about:
   a. -65.2°F. (-54°C.)
   b. -45°F. (-43°C.)
   c. +32°F. (0°C.)
   d. +300 to 400°F. (149 to 204°C.)
2. The boiling point of ethyl amine is:
   a. -40°F. (-40°C.)
   b. 0°F. (-18°C.)
   c. +62°F. (17°C.)
   d. +723°F. (384°C.)
3. The ignition temperature of ethyl ether is:
   a. -49°F. (-45°C.)
   b. +95°F. (35°C.)
   c. +356°F. (180°C.)
   d. +1000°F. (538°C.)
4. Most ethers have a tendency to form explosive peroxides. This is particularly true in the case of:
   a. Isopropyl ether
   b. Ethyl ether
   c. Butyl ether
   d. Divinyl ether

5. Forming emulsions with a coarse spray can be an effective method of extinguishing fires in:
   a. Kerosene
   b. Alcohols and ketones
   c. Aldehydes
   d. Heavy oils
6. One of the problems of naphtha is:
   a. High toxicity
   b. It will explode spontaneously
   c. Wide flammable range
   d. Several different kinds, with different flash points.
7. The most toxic common hydrocarbon is:
   a. Gasoline
   b. Benzene
   c. Turpentine
   d. Hextane
8. Which of these liquids is *not* subject to spontaneous heating:
   a. Linseed Oil
   b. Gasoline
   c. Turpentine
   d. Cottonseed Oil

FURTHER READING

1. *Flammable and Combustible Liquids Code Handbook,* National Fire Protection Association, 1 Batterymarch Park, Quincy, MA., 1987.

VISUAL AIDS

American Heat Video Production, 240 Sovereign Ct. Ste. C, St. Louis, MO. Paraffin Wax Company Fire. Oshkosh, Wisconsin, Vol. 7, Program 7. Isopropyl Alcohol Fire. Tampa Florida, Vol. 5, Program 5. Inks and Rosins Fire. Cincinnati, Ohio, Vol. 5, Program 3.

Petroleum Fire! *The National Fire Protection Association,* 1 Batterymarch Park, Quincy, MA.

Foam. Detrick Lawrence Corporation. The Emergency Film Group, 225 Water St., Plymouth, MA 02360 (28 minutes, VHS format).

Hazchem 7: Benzene, Toluene, and Xylene. Detrick Lawrence Corporation, The Emergency Film Group, 225 Water St., Plymouth MA. (28 minutes, VHS format).

Flammable Liquids and Emergency Response: Proper Recognition and Safe Handling, Chemical Manufacturer Association, Lending Library; 2501 M St., NW; Washington, DC.

Flammable Liquids in Bulk

4

Construction requirements for flammable liquid tanks, pipings, valves and fittings, and their installation and location is an enormous subject. The storage code for flammable liquids forms a major part of any fire prevention course. We will give particular attention to the problems firefighters encounter when fires and explosions occur in bulk flammable liquid storage, in tanks, tank trucks, warehouses, and underground pipelines.

Why do flammable liquids figure so prominently in annual reviews of the biggest fires? Not only because of their inherent hazard, but also because of the enormous quantities involved in storage and shipping. A single large tank may contain 200,000 barrels of fluid or more. A tank farm or refinery may have several dozen such tanks, each a potential headline. Flammable liquid drum storage may be concentrated in a yard or warehouse, with one 55-gallon drum piled atop another and hundreds in a small area. A tank truck can carry thousands of gallons of gasoline along a busy street for delivery to the service station on the corner.

Fortunately, bulk flammable liquid fires are rare, and a fire department is seldom called upon to face such an emergency. The best way to prepare for such incidents is through examination of the victories and defeats of other fire fighters. One of the most valuable services provided by the NFPA, (National Fire Protection Association), NTSB (National Transportation Safety Board) and Fire Engineering Magazine is their thorough investigation and reporting of large-loss fires. These reports deserve careful study. (See the bibliography at the end of this chapter.)

TANK FIRES

Much of this section states the obvious, but overlooking the obvious is all too common at the scene of a large flammable liquid fire. The sizeup of such a fire can be extraordinarily difficult. Radiated heat may make close approach impossible; smoke may obscure vision. Yet the firefighter must correctly evaluate a great many factors when facing a tank fire.

The Water Situation

Adequate pre-fire planning is a must if one is to identify the location of useful hydrants and the size of the mains that feed them. In several bulk fire histories, hydrants were in poor position, requiring long, awkward, time-consuming hoselays. In other cases, water supplies were insufficient. Make no mistake about it—the amount of water available is vital to firefighting operations. A great deal of heat may have to be dissipated before firefighters can approach within working distances. Only the use of protective water curtains will allow access to necessary valves. Cooling water sprays *must* be applied to maintain the integrity of exposed tanks. In some cases, the volume of water used for cooling has literally caused flammable liquid tanks to float on the water. The fire officer must guard against this by ensuring proper water drainage. If there is not enough water to use for protecting exposures, the evacuation of firefighters to safe distances—at least 2,000 feet (600 meters)—is required.

Liquids Involved

The situation of a tank fire is where knowledge of flammable liquids is put to work. What is their specific gravity or water solubility? It may be possible to blanket, emulsify, or dilute them. If their flash point is in the right range, direct application of a fog pattern may bring quick extinguishment. Knowledge of flammable liquids will also increase the accuracy of expected outcomes. Proper type of foam to use can be determined. Waste of time or resources in a futile effort to extinguish a large gasoline fire with water will be minimized. The possibility of steam explosions and boilovers can be determined and resolved effectively.

At this time, examination of one type of boilover in more detail is necessary. **Crude oil**, as stated earlier, is a mixture of petroleum fractions. Before it is refined, crude often contains a significant amount of water, has a wide range of boiling points, and is extremely viscous. In a storage tank, the water eventually separates and settles at the bottom. After a crude oil tank is on fire for several hours, the lighter fractions burn off, and hot, heavier fractions sink toward the bottom of the tank in what is called a "heat wave." This downward progress of a heat wave can be observed by the use of a strip of heat-detecting paint or by watching the action of water flowing down the sides of the tank; water is likely to vaporize when it reaches the area of the heat wave. Heat waves travel at a varying rate, approximately one to four feet per hour. When they reach a point five feet (1 1/2 meters) above the known level of the bottom water, it is time to leave the scene. The hot fuel converts the water to steam that becomes trapped by the heavy, viscous crude oil. The expanded contents of the tank are violently expelled. A crude oil boilover is signaled by an increase in flame intensity. It is highly unlikely that dikes will contain the boilover unless they are especially constructed for that purpose. Finally, don't be in a hurry to return; successive boilovers in crude oil tanks are not uncommon.

The following incident in Milford, England is a classic example of the events leading up to and including a boilover sequence.

At 11:00 A.M. on August 30, 1983 a fire occurred in a 250-foot diameter crude oil tank approximately half full. By 12:00 noon, crude oil began to be removed from the tank through internal piping. At first it was believed by firefighting personnel that the fire was seal fire around the edge of a floating roof tank. Unknown to the firefighters, the floating roof tank had developed cracks which allowed crude oil to seep through the roof and pool on top. A week previous to this fire the area had experienced heavy winds and torrential downpours. Water collected on the roof. It was estimated that approximately 250 tons of water and oil had accumulated on top of the floating roof in the weeks prior to the fire. Since the initial assessment of the incident perceived this to be a seal fire, foam application was initiated. This tactic was used for the first hour of the fire. It was estimated that by approximately 3:00 P.M. 700 tons of water and crude oil had accumulated on the roof of the tank. At approximately 4:30 P.M. an enormous increase in fire activity occurred resulting from what most believe to be the sinking of the floating roof. At this time it was determined that the entire surface of the crude oil was on fire. Upper portions of the tank began to buckle inward as a result of the intense heat. A cooling action operation was employed until enough equipment and resources could be amassed for an all-out foam attack on surface of the liquid. It was estimated that approximately 56,000 gallons of foam material would be required to fully extinguish the fire. It would take several hours to gather the required foam from surrounding agencies, therefore the foam operation was scheduled for 12:30 A.M.. At 12:30, just before the foam application was about to commence, a boilover occurred without warning. It was estimated that oil was ejected over 3,000 feet in the air and filled the holding dike to two-thirds capacity. The fire ball was so bright that it caused street lights to turn off in a town several miles away. Grass fields in the surrounding areas erupted in flames from radiated heat. After re-establishing firelines over the next few hours, a second boilover occurred as violently as the first. Tanks located in close proximity to the tank on fire had been drained as a precautionary measure. Due to the radiated heat of the second boilover the vast vapor space inside the adjacent tanks ignited and an explosion and tearing of the tank seams was experienced. Eventually the fire was brought under control without further incident.

The Physical Layout

Preplanning will develop a detailed picture of the physical layout. Are there dikes to confine a spill fire to a localized area? If so, only one tank, or relatively few tanks, will be immediately involved. Since preventing spread is the primary consideration in any flammable liquid fire, dikes solve the problem of confinement, although another problem immediately takes its place: Each of the tanks in the involved area will be subjected to a great deal of stress as the fire continues to burn around them. Using enough water to protect the tanks can overflow the dikes, destroying their value. Can these burning liquids be safely drained away into impounding basins where, away from exposures, they can burn out? Only the value of the liquid, often the cheapest part of the potential property loss, will then be lost.

Types of Tanks and Piping

Preplanning can also tell you the types of tanks in the occupancy and the location of piping, vital bits of information. As Battalion Chief Erven of the Los Angeles Fire Department has commented, "When fighting tank fires, often the only method of extinguishment is to pump out the contents of the tanks. This is why modern tank farms place valves outside the dikes. In one large tank fire at Union Oil in Wilmington, thousands of barrels of gasoline from the burning tanks were pumped into empty tanks 15 miles away." Most tanks have vents designed to prevent vapor overpressures.* You should know if these vents are sufficient only for normal "breathing" of the tank. Inadequate vents, such as these, are incapable of relieving the tremendous increase in vapor pressure caused by a surrounding spill fire. Tanks vented in this way may explode like overheated boilers, raining down thousands of gallons of burning liquid. The increasing sound of vapors escaping from the vents will indicate the buildup of pressure in an overheated tank. This sound has been variously described as "a whistle," "a jet engine," and so forth. Unless water can be immediately applied to such a tank to cool it and lower the interior pressure, evacuate the area. The first firefighting efforts should be devoted to extinguishing any ground fire that is heating the tank. When there is a fire at the vent, a cooling spray must be maintained over the area involved to prevent localized overheating, which sometimes is followed by a violent rupture of the entire tank. These types of ruptures are commonly experienced in large flammable liquid tank fires and with some warning of the eminent failure.

| Horizontal | Cone-type | Floating Roof | Spheroid |

FIGURE 4–1. Examples of tank design.

* Spheroidal and cone-type tanks also have vents for relief of overpressures.

However, they can happen when and where they are least expected. On April 4, 1987, the Bradford Township Volunteer Fire Department of Bradford, Pennsylvania, responded to a house fire that resulted in the violent explosion of a 250-gallon fuel oil tank in the basement of a residential home. Firefighters must be aware of the critical factors present in every situation in order to prevent disastrous outcomes. In Bradford Township, the township chief recognized the potential for disaster and reassigned hoselines in the basement just prior to the explosion, with the result that firefighters in the basement sustained only minor injuries.

Although cooling of the tanks in a flammable liquid fire is an acceptable practice, there exists some evidence that there may be exceptions to the rule. Miles Woodworth, NFPA flammable liquids engineer, points out some possible exceptions in an article in *Fireman's Magazine*, July 1956.

> A fire that is burning at a tank vent with a yellow-orange flame, emitting black smoke, indicates a vapor-rich condition within the tank that is above the flammable or explosive limits. This type can be extinguished by smothering with steam, dry powder, CO_2, or wet blankets. There is no danger of an explosion.
>
> Fire at a tank vent that burns with a snapping, blue-red, nearly smokeless flame indicates a vapor-air mixture within the tank that may be explosive. There is constant danger of an explosion should the flame reach the inside of the tank. The best attack procedure, therefore, is to maintain a "positive pressure" within the tank by pumping liquid into it. When a vapor-rich condition is indicated by change in the flame character to a smoky yellow-orange flame, danger of explosion has passed. Extinguishment can then be accomplished as stated in the previous paragraph.
>
> Never apply cooling water to tanks burning with a snapping, blue-red, nearly smokeless flame at the vent. Nor should such tanks be pumped out. Several of the tank vent fires reported to the NFPA featured injury to firefighters when water applied to cool the tank shell caused an explosion.

This reinforces the adage that nothing is for sure. There are situations where you are damned if you do, damned if you don't. Sometimes, there is just no completely acceptable solution to the problems posed by a bulk flammable liquid fire. However, remember, "When in doubt, get out!"

BLEVEs

Boiling Liquid Expanding Vapor Explosions—BLEVEs—can occur whenever flammable liquids are rapidly heated to temperatures well above their boiling points.

The resulting expansion of the vapors causes internal pressures beyond the tank venting mechanism capabilities, followed by the inability of the tank structure itself to withstand the high internal pressures. This combination of factors can result in an instantaneous release and ignition of the flammable vapors caused by structural failure of the tank. An explosion, called a BLEVE, then occurs. There is often a time delay prior to the explosion—a factor that can be fatal, for the time delay allows both firefighting personnel and equipment, as well as spectators, to approach close enough to become potential casualties of the extremely violent explosion. A BLEVE is usually accompanied by a huge fireball.

A BLEVE is not limited to stationary storage tanks but can involve railroad cars and tank trucks as well. It can be destructive to life and property over a radius of more than 2,000 feet (600 meters).

The approach to suppression, or in these instances *prevention*, of a BLEVE is immediately to place very heavy stream appliances into operation. They should be positioned where the water can be played on the upper sides and top of the tanks involved. If at all possible, these heavy streams should have at least a 500-gallon-per-minute (2000-liter-per-minute) capacity and should be unstaffed.

Horizontal Tanks.   Horizontal tanks, larger versions of the familiar railroad tank car, have their own peculiar way of reacting to excessive pressures. They may explode like other tanks, but more often they sky-rocket. One end of the tank gives way, and the tank rockets off its supports in response to the same principle that sends a missile into orbit from its launching pad at Cape Canaveral. A fundamental rule in tank firefighting is *never approach a horizontal tank involved in a fire from either end.* Always approach it from the side.

Unfortunately, this rule was not so clearly established before a fire in the Midwest a decade or so ago. *After* the fire, of course, everybody knew the rule. In any event, firefighters were stationed at the ends of several horizontal tanks, fighting a severe flammable liquid spill fire. Intent upon confining the fire to a small area, they used hose lines to sweep the burning gasoline back under the tanks instead of flushing it into an open area. The vents on this group of tanks were too small to relieve the tremendous overpressures that resulted. As a result, one tank rocketed directly at a group of firefighters ninety feet away, killing six of them. (Horizontal tanks can fail still another way. If they sit upon unprotected steel supports, instead of concrete or masonry, the supports can buckle quickly when involved in a fire. The tank hits the ground or the piping breaks. Either way, the contents are released and an explosion often results.)

Floating Roof Tanks.   As their name implies, floating roof tanks eliminate a potentially explosive vapor space by having the roof rest directly upon the liquid contents. Instead of vents, this type of tank is built with a deliberately weakened roof seam. Flammable liquid fires often begin with an explosion, and a tank with its roof ruptured or missing is not subject to a pressure explosion. But when firefighters reach it, it will probably be on fire, and heat will cause the sides of the tank to fold in a few feet above the level of the liquid.

The best way to extinguish an **open tank fire** is through the application of the proper foam blanket. *No extinguishing agent is as successful as foam on large flammable liquid tank fires.* When applying cooling streams of water on a tank equipped with a floating roof that is exposed by an adjacent fire, firefighters must be careful not to exceed the capacity of the scuppers. At a marine oil terminal fire, a fireboat applied such a large quantity of water to prevent a 60,000-barrel tank of gasoline from being involved that the tank top became overloaded and sank to the bottom. This resulted in a large open tank of gasoline that had to be protected from ignition by the maintenance of a thick blanket of foam.

Perhaps the only fairly safe flammable liquid tanks are those two feet underground. Such tanks are almost impervious to a fire above them. If there are no exposed or exploding tanks to hold your attention, you should give consideration to the piping. Involved pipes may fail and release the entire contents of a tank into a running spill fire, thus creating a fire that is extremely difficult to combat. A foam blanket underneath vulnerable piping may help control the situation until the proper valves can shut off the flow.

Plant Manager

Get the plant manager's advice if you can. Company representatives can provide invaluable information as to location of valves, dike drains, level of bottom water in crude oil tanks, and so on. One method of fighting storage tank fires is the subsurface injection of aerated foam into the tank. The foam rises to the top of the tank and provides a layer of foam between the liquid and the exposed fire. The plant manager can provide the location of valves and piping that provide access to the tanks. Additionally, in the case of a piping fire at the lower levels of the tank, application of the specific gravity principles of flammable liquids would allow for the subsurface pumping of water into the tank to raise the water level in the tank to a point above the level of the leaking pipes. At this point, water would be leaking from the damaged piping, and the flammable liquid would be on top. If the plant manager is not available, there is usually at least one person on site to provide needed information. Don't be hesitant to ask company individuals for advice; they work in that environment every day, and their insight is usually helpful.

Keep the Public Away!

Have the police department keep spectators and all persons other than firefighters back at least 2,000 feet (600 meters). This is most important. If a large tank lets go, the amount of heat released is overwhelming. You have enough of a problem protecting firefighters from this catastrophe without worrying about the public. Although this may seem to be unneeded advice, the history of bulk flammable liquid fires has proven their crowd appeal and the danger to onlookers. Keep them away even though it will spoil their photographs.

DRUM STORAGE

Let's face a few facts about flammable liquid drum storage. Rules and recommendations for safe handling and storage are more often ignored than followed. Companies tend to use every possible inch of storage space. As a result, the number of drums in a storage area will exceed requirements for safe size and height of piles, and for the permissible quantity within a particular area. The maximum number of drums allowed in a particular sprinklered area depends upon the location of the storage space and whether it is detached, cut off, or part of the main building. The type of liquids stored is also a factor. (For an excellent summary of this entire question, see "Fire Protection Guide to Hazardous Materials," published by NFPA.)

Flammable liquid storage is not always strictly segregated or in detached, noncombustible buildings with dispensing done in cutoff, properly arranged, and protected rooms. Ignition sources can be present in both areas. Housekeeping may be substandard with no inspection program for leakage. Drainage systems can be inadequate; sprinkler systems, nonexistent. Most of the firefighters with experience in industrial inspection know that not all companies strictly observe safety practices. Some do; others are very haphazard and negligent. It can be a long time between fire department inspections. This is another reason why flammable liquids cause an appreciable percentage of industrial fires.

In an unopened drum of good construction, a flammable liquid is only a potential hazard. Nothing will happen as long as it is safely tucked away. Trouble begins when the liquid gets loose in some way: by leakage caused through rough handling or falling from a high pile; deterioration of the container; unsafe dispensing practices; structural failure of the drum because of heat or a nearby explosion. The amount of trouble, as has been experienced, will depend on the type of liquid set free, potential ignition sources, and type of built-in protective equipment.

Once again, the particular properties of a given liquid (see Chapters 1 through 3) determine its degree of hazard. When a liquid is liberated, for whatever reason, it must do what its properties dictate. Vapors will travel according to their densities, and liquids will seek their level. If a vapor cloud within the flammable range encounters an ignition source above the ignition point, an explosion *always* occurs.

Companies that store drums of flammable liquids can reduce the risk of fire and explosion in a number of ways. Good ventilation of rooms where flammable liquids are stored or handled is vital. Ignition sources should be eliminated wherever possible: Open flames, sparks, electrical equipment, static, hot surfaces, radiant heat, friction, overheating, spontaneous ignition, and all other sources that might not be considered until it is too late.

But even when the best practices are followed, including safety precautions, trouble can and does start. When it does, the most efficient safeguard in areas where flammable liquid drums are kept is the automatic sprinkler system. Of course, sprinklers have limitations. The amount of drum storage in a room can exceed sprinkler control capability.

Explosions can damage the system, knocking it out. A small leak from a drum can provide fuel for a localized fire intense enough to rupture nearby drums, but not big or hot enough to activate the sprinkler system. The flash from one bursting drum can open hundreds of sprinklers, some well beyond the initial fire area, and severely overtax the system. Finally, sprinklers will not extinguish many low-flash-point flammable liquids, and flowing water may carry burning liquid, spreading the fire to uninvolved areas. Nevertheless, a sprinkler system can prevent a disaster. It will cool surrounding drums and exposures. With proper drainage, it can flush away the burning contents of a drum before exposed drums are heated dangerously. Without sprinklers, the heat of a spill fire can cause a pile of drums to explode like a pan of popcorn.

Upon arrival, firefighters complete the work sprinklers have begun. A firefighter has one big advantage over a sprinkler system: He or she can think. The firefighter makes sure the system is not turned off prematurely, supports it through fire department connections, then sets about confining the fire and using the proper extinguishing agents.

TANK TRUCKS

A truck driver is not a firefighter and should not be expected to know the contents of tanks unless he or she is delivering gasoline. Frequently, the driver neither knows nor cares, but he or she does have bills of lading that will tell us exactly what liquid is in the tanks. The driver can be of great value to fire fighters because such a person usually knows the operation of the truck and is familiar with the location of valves.

All firefighters should take a good look at a tank truck, if they have not already done so. They are beautifully designed to reduce the possibility of a fire. Internal valves will close to prevent the drainage of the tank if piping breaks and can be operated remotely. If the vents on a modern tank truck are free to operate (and they will, if the truck is not overturned), they will take care of overpressures. There has been no record of a tank truck's rupturing or exploding for this reason. Any explosions that occur come from the tires, the truck's own fuel tanks, or sudden ignition of spillage.

Some tanks are made of steel and others of aluminum. Steel tanks, of course, are stronger and less likely to break open if the truck overturns. Aluminum tanks will melt far more rapidly if the tank is involved in a spill fire. This is not as dangerous as it sounds. It presents us with the relatively simple problem of several small, open tank fires as the aluminum melts down just above the liquid level to reveal the internal compartments of the tank. These small fires can quickly be extinguished by foam or dry chemical, supported by fog streams for cooling and prevention of re-ignition.

If the truck is overturned, running-liquid spill fires can be controlled by hose streams and directed to safe areas for burnout. A spill that is *not* on fire is more complicated. Depending on the severity of the spill, it may be necessary to shut off ignition sources or even evacuate downwind areas. Foam blankets can be employed. Vacuum tank trucks, such as those used in the oil fields or those used to pump out septic tanks, will quickly and effectively suck up flammable liquid spills and pools.

SERVICE STATIONS

Gasoline tank trucks are generally heading for a neighborhood service station. With the exception of car fires, this type of occupancy constitutes our greatest involvement with flammable liquids. There are over 200,000 service stations in the United States. Each of them sits above at least one gasoline tank. Although some of the following safety precautions are impossible to enforce, they are directed at the causes of most service station fires.

1. Do not permit smoking when gasoline is being dispensed. Post clear signs to that effect.
2. Motorists should turn off their ignitions when their tanks are being filled.
3. A competent individual should be in the vicinity of the vehicle being filled, even though automatic nozzles are being used.
4. Sale of gasoline in glass jugs or bottles is forbidden.
5. When tank trucks are making a delivery, they should be on service station property, not on the street.
6. Waste-flammable liquids must not be dumped into sewers.
7. Open flames are forbidden.

From a firefighting standpoint, the most important requirement for service stations is a clearly identified switch, readily accessible, that cuts off electric power to the pumps in the event of a fire or physical damage to the dispensing units. Our strategy should include the immediate closing of this switch to prevent a malfunction of the pumps, causing an increase in the amount of gasoline involved. Beyond this, the primary consideration remains the same: Confine the problem. Protect exposures; don't wash gasoline toward them. Let the fire burn out if it cannot be quickly extinguished by portable foam or dry chemical extinguishers. If there is an extensive spill fire, with resulting danger to exposures, sterner measures must be employed, such as coordinated 1 1/2" fog lines. Directed with intelligence, fog lines can surround and direct the course of the spill, even though they will not extinguish it.

HIGH-PRESSURE PIPELINES

High-pressure pipelines are the preferred mode of transportation of flammable liquids between flammable liquid storage sites. High-pressure pipelines are capable of delivering immense quantities of bulk liquid in a fraction of the time it takes to transport it over the highway. Almost all flammable liquid deliveries to bulk storage facilities are transported through underground high-pressure pipelines. Pipelines usually run alongside railroad tracks because of the ease of access and the fact that pipelines were originally owned by railroad companies. Because the location is in close proximity to railroad tracks, damage to the pipeline should be a major concern in any train derailment. Such an incident happened in San Bernardino, California, in May 1989. On May 12, a train derailment caused the closure of a high-pressure pipeline that connected the high desert areas of southern California to the Los Angeles basin. The location of the incident was at the base of the San Bernardino Mountains in the city of San Bernardino. The pipeline remained closed for 4 days while inspection of the pipeline and repairs to the railroad track could be accomplished. On May 16, the pipeline was placed back in operation. On May 25, a loud noise was heard by the residents in the vicinity of the original train derailment. Residents came out of their houses to find gasoline running off the roofs of their houses like rain. Shortly after, the gasoline vapors found an ignition source, a three-block area was instantly turned into a mass of burning liquid and houses. Upon arrival of the first fire engines, 10 houses were totally involved in fire. See photo 4-1. The pipeline company was immediately notified, and the line was shut down. However, because the pipeline travels over the San Bernardino Mountains at an elevation of almost 5,000 feet, gasoline continued to flow for an additional three to four hours. Most pipelines are constructed with check valves to eliminate backflow. In this situation the check valves failed, and eight miles of pipeline, with an extremely high head pressure resulted in a pressurized column of flammable liquid that extended 500 feet in the air, spewing fire and destruction of the residential area adjacent to the tracks. It was estimated that during the five hours the fire burned, 400,000 gallons of gasoline was consumed. Surprisingly, no one was killed. However thirty-one civilians were injured, five critically.

There are incidents on record of high-pressure pipelines' rupturing in highly populated areas, the most notable being an incident on the city border of Los Angeles and Culver City. This incident was caused by a trenching operation not involving the high-pressure pipeline company. The trencher operator was unaware of the existence of a high-pressure pipeline.

High-pressure pipelines carrying a great variety of flammable liquids (as well as toxic and flammable gases) pass beneath city streets today. Do you

PHOTO 4–1. Fire from high-pressure gasoline pipeline rupture. (Photo by San Bernardino Fire Dept.)

know where they are located? Do you know what they transport? Do you have a plan of action in the event of an accident?

Even an eight-inch pipeline can provide you with a flammable liquid—*in bulk.*

SUMMARY

Flammable liquids are a necessity for today's lifestyle. Almost everything produced or consumed in today's society has, at one time or another, relied on a flammable liquid for either its production or transportation to the marketplace. If electricity is required, most likely a flammable liquid is involved. Therefore, emergencies involving flammable liquids are inevitable. For the most part, flammable liquids, at the point of use, do not pose as big a threat of becoming involved in an incident as when they are stored or transported in bulk, to the point of use site. The large loss fires, both in dollars and lives, involving flammable liquids have been and will always be where they are stored or transported in bulk. It is incumbent upon firefighting personnel to become extremely familiar with the special circumstances of flammable liquid bulk storage and transportation in their community.

REVIEW QUESTIONS

1. When evacuation of firefighting personnel at an oil tank fire is required, the *minimum* safe distance is:
   a. 100 feet (30 meters)
   b. Beyond the smoke
   c. 2000 feet (600 meters)
2. A crude oil "boilover" is signaled by:
   a. An increase in black smoke
   b. Jets of flaming oil
   c. An increase in flame intensity
3. The primary consideration in controlling a flammable liquid fire is:
   a. Using fog
   b. Immediate ventilating
   c. Preventing spread
   d. Toxic vapors
4. A snapping, red-blue, nearly smokeless flame, burning at an oil-tank vent, shows us that there is:
   a. No danger of explosion
   b. A vapor-rich condition in the tank
   c. Great danger of explosion
5. Never approach a horizontal flammable liquid tank that is on fire:
   a. From the sides
   b. Without using fog
   c. Without contacting the plant manager
   d. From the ends.

BIBLIOGRAPHY

1. *Flammable and Combustible Liquids Code Handbook,* National Fire Protection Association, 1 Batterymarch Park, Quincy, MA, 1987.
2. "On the Job in California: Train Derailment and Pipeline Explosion Devastate San Bernardino," *Firehouse*, October, June, 1991.
3. "Over the Top: Techniques and Logistics for Extinguishing Large Tank Fires," *Industrial Fire Safety,* November-December, 1992.
4. "BLEVE! Basement Fuel Oil Tank Explodes; Two Pennsylvania Firefighters Escape," *Firehouse*, April, 1989.
5. "Flammable Liquid Drainage and Containment," *Record*, July-August, 1992.

FUTHER READING

1. Annual Report on Hazardous Materials Transportation," U.S. Department of Commerce; National Technical Information Service (NTIS), 5285 Port Royal Road, Springfield, VA 22161; published annually.

VISUAL AIDS

Boilover—ARCO Tank Fire, Milford, England, 8/30/83, Hazardous Materials Library, FEMA Region IX, Presidio of San Francisco, FL5-3-284A, VHS format.

Fighting Petroleum Storage Fires. National Fire Protection Association, 1 Batterymarch Park; Quincy, MA 02169. #BU-FL-55, 16mm film.

Gasoline Tank Truck Emergencies. National Fire Protection Association, 1 Batterymarch Park; Quincy, MA 02169. #BU-FL-71, 16mm film.

Pressurized Gases

Solids retain their size and shape. Liquids have a definite size and at least one shape, a flat top surface. But gases have neither—no size, no shape, and, seemingly, no rhyme nor reason.

GENERAL PRINCIPLES

Gases will expand upon heating (indefinitely, if you let them) and contract upon cooling. They are completely elastic and will fill any container. Furthermore, they will not settle in it. If you put one liter of liquid in a two-liter container, one liter of liquid will flow to the bottom; put one quart of gas in a two-quart container and you have two quarts of gas. The gas pressure will be the same at the top as it is at the bottom.

Different gases will mix completely. If you put a lightweight gas and a heavyweight gas in the same container, there will not be the sharp line of division you would find in a bottle of mixed gasoline and water. Even gases with different vapor densities will not separate completely. A small amount of the lightweight gas will be found at the bottom of the container, some of the heavy gas at the top.

The only logical explanation for the behavior of gases is that they are composed of tiny particles in constant random motion. The higher the temperature, the more violent the motion, the greater the urge to expand, to fly off in all directions. A great many gas "laws" have been suggested to predict this behavior of gases. Many of them, however, are not easily applied on the fire ground.*

*Avogadro's Law says that equal volumes of *any* gas contain the same number of molecules. Henry's Law and Roualt's Law have to do with solubility of gases. Dalton's Law predicts the total pressure of a mixture of gases. Graham's Law is a description of the rates of diffusion of the gases. If interested, consult a physics textbook for more information.

PHOTO 5–1. **DOT nonflammable gas labels and placards.** This is the DOT label (green with black lettering) required on cylinders when compressed, nonflammable gases are being shipped.

However, it is valuable to know that the **volume** of a gas (generally the size of the container in which a gas is stored), the **temperature** of a gas, its **pressure,** and sometimes its **amount** are all interrelated. Change the value of one of these factors, and the value of another or perhaps all of them are changed.

GAS PRESSURE

Of primary interest to firefighters is the pressure of a gas. This is where its danger might lie. How can the pressure of a gas be increased? Pressure is caused by the impact of gas molecules against the sides of a container that prevents them from heading for parts unknown. The more impact in a given amount of time, the higher the pressure. The pressure of a gas can be increased by increasing the number of impacts. There are three ways of accomplishing this, each involves changing one of the other factors we mentioned.

1. *Decrease the volume.* A smaller container will shorten the distance a gas molecule has to travel before it contacts the sides. This will increase the number of collisions within a given time period—there-

PHOTO 5–2. DOT nonflammable gas labels and placards (cont.). The nonflammable gas placard must be green with white symbol & inscription.

fore, more pressure. The amount of pressure is predicted exactly by Boyle's Law. Cutting the volume in half, when the temperature and amount of gas are held constant, will double the pressure. (Boyle's Law reverses this wording—"At a given temperature, the volume occupied by a gas is inversely proportional to the pressure"—but means the same thing.) (See Figure 5-1.)

2. *Increase the amount of gas within the cylinder.* More gas means more molecules to collide against the cylinder walls.

3. *Heat the gas.* A hot gas has faster-moving molecules. They will strike the sides of the cylinder more often than slower-moving molecules.

Combined Gas Law

The amount of pressure increase coming from a rise in temperature can be predicted by the Combined Gas Law (a combination of Boyle's and Charles's laws). This law can be extremely useful. Most gas cylinders will release their contents through pressure-relief devices when the pressures within them reach such a level that the cylinder is in danger of bursting. Whether the released gas is flammable or toxic can be an important factor in a fire situation. An approaching fire can quickly heat a cylinder to 500, 1,000, or even 1,500 degrees Fahrenheit.

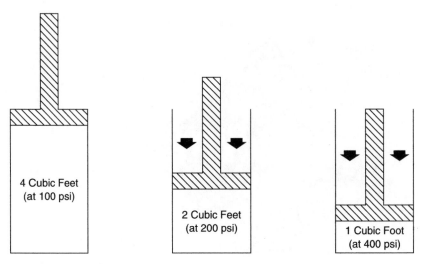

FIGURE 5–1. Volume-pressure relationship.

The Combined Gas Law says, in effect, that the pressure of a gas is dependent upon the temperature to which it is exposed and the volume of the container it is in. Because the volume of a steel cylinder expands very little, this part of the problem can be ignored. The interest lies in the interrelationship of temperature and pressure. Temperature means absolute temperature: the thermometer reading plus 459.7°F (273°C). (Absolute zero is –459.7°F (–273°C), often given as –460°F.) Pressure means absolute pressure, which includes atmospheric pressure.

What would happen to the interior pressure of a gas cylinder at 1,000 psi at 40°F if it were heated to 1540°F? Forty plus 460 is 500 degrees absolute, 1540 plus 460 is 2000 degrees absolute.

Minus the constant volume factor, the simplified equation looks like this:

$$\frac{\text{Temperature 1}}{\text{Temperature 2}} \text{ is in ratio to } \frac{\text{Pressure 1}}{\text{Pressure 2}} \text{ or } \frac{500}{2000} : \frac{1000}{????}$$

The obvious answer is 4,000 pounds of pressure within the cylinder, enough to burst it. Many equations would be more difficult, but no more difficult than some problems in hydraulics. There is even a quick rule of thumb to go by: *Beginning with a gas at normal temperatures, an increase of 500 degrees will just about double its pressure; an increase of 1,000 degrees will triple its pressure, and 1,500 degrees will quadruple it.* In Celsius (metric) measurement, an increase of 273°C will double the pressure.

Raising the temperature of a gas raises its pressure; lowering the temperature lessens the pressure, and vice versa. Raising and lowering the pressure of a gas raises and lowers its temperature. Many firefighters are

PHOTO 5–3. DOT nonflammable gas labels and placards (cont.). The oxygen placard must be yellow with a 1/2-inch (12.7-mm) white border. The symbol and inscription must be black. Although nonflammable, the yellow background denotes an oxidizing agent which supports combustion.

familiar with this phenomenon from filling oxygen or compressed air bottles. This principle is what makes refrigeration systems work.

FIREFIGHTING PRESSURIZED GASES

Industrial gases are shipped and stored in two ways: as liquids that generate a stabilizing vapor pressure, or as gases in high-pressure cylinders of various kinds. Certain hazards, common to all gases stored under high pressure, make it convenient to discuss them as a group. This chapter concentrates on them, although it must be remembered that all arbitrary divisions such as this contain exceptions. Many gases are shipped and stored in both forms.

High-pressure gases are universally valuable: Methane is one of our most important fuels; acetylene and hydrogen are used in welding and cutting operations; oxygen has many medical and industrial uses; ethylene is used both as an anesthetic and to ripen fruit. Gases are used for such diverse purposes as rocket propellants, fumigants, refrigerants, and insecticides. Table 5–1 lists the gases to be examined in this chapter, but this is by no means a complete list of the pressurized gases employed by U.S. industry. Gases with special properties or uses will be part of later

TABLE 5-1. Pressurized Industrial Gases

	Lower Flammable Limit (% by vol. in air)	Upper Flammable Limit (% by vol. in air)	Ignition Temp.	Vapor Density (air = 1.0)	Boiling Point	Water Soluble	Typical Cylinder Pressure (PSI)	Color of Required DOT label
Acetulene (C_2H_2) (ethyne; ethine)	2.5	100.0	581°F 305°C	0.91	-118°F -83°C	No	250	Red
Air (mixture)	(Nonflammable; will support combustion.)			1.0	-317°F -194°C		2,000	Green
Argon (Air)	(Nonflammable; will not support combustion.)			1.4	-302°F -186°C	Slightly	2,000	Green
Carbon dioxide (CO_2) (dryice)	(Nonflammable; rarely supports combustion.)			1.5	-109°F -78°C (sublimes)	Yes	830	Green
Cyclopropane (C_3H_6) (trimethylene)	2.4	10.4	928°F 498°C	1.5	-29°F -34°C	No	75	Red
Ethane (C_2H_6)	3.0	12.5	882°F 472°C	1.00	-128°F -89°C	No	530	Red
Ethylene (C_2H_4) (ethene)	2.7	36.0	842°F 450°C	1.00	-155°F -104°C	Yes	1,200	Red

TABLE 5-1. Pressurized Industrial Gases (cont.)

	Lower Flammable Limit (% by vol. in air)	Upper Flammable Limit (% by vol. in air)	Ignition Temp.	Vapor Density (air = 1.0)	Boiling Point	Water Soluble	Typical Cylinder Pressure (PSI)	Color of Required DOT label
Helium (He)	(Nonflammable; will not support combustion.)			0.14	-452°F -269°C	Very slightly	2,000	Green
Hydrogen (H$_2$)	4.0	75.0	932°F 500°C	0.1	-422°F -252°C	Slightly	2,000	Red
Methane (CH4) (marsh gas; cooking gas)	5.0	15.0	999°F 537°C	0.60	-269°F -162°C	No	2,000	Red
Nitrogen (N$_2$)	(Nonflammable; rarely supports combustion.)			0.96	-320°F -196°C	Yes	2,200	Green
Nitrous Oxide (N$_2$0) (laughing gas)	(Nonflammable; will support combustion.)			1.52	-129°F -89°C	Yes	800	Green
Oxygen (O$_2$)	(Nonflammable; will support combustion.)			1.1	-267°F -183°C	Slightly	2,200	Green

chapters, and Chapter 6 will cover liquified gases, including ammonia and the LP (liquified petroleum) gases.

Several questions must be answered before firefighting operations on pressurized gas fires can be successful.

Type of Gas Involved

Learning the type of gas involved must be the first objective. Plant personnel may identify a gas, or identification may come from names or formulas on containers and bills of lading. The color and shape of the gas cylinder itself are of doubtful value in identification. Manufacturers tend to use their own color-coding system. These vary. Some manufacturers store all their gases in cylinders of the same color and use them interchangeably. There have been discussions that the Compressed Gas Association will attempt to standardize cylinder colors throughout the industry. This would be of great value. Until this happens, the colors listed in Table 5–2 are recommended by the National Bureau of Standards (and sometimes adhered to).

The Department of Transportation (DOT) requires certain labels when cylinders of compressed gas are shipped between states. This labeling

TABLE 5-2.   Identifying Colors of Gas Containers

Color	Gas
Green	Oxygen
Light blue	Nitrous oxide
Brown	Helium
Orange[a]	Cyclopropane
Brown and green	Helium and oxygen
Red	Ethylene or hydrogen
Gray	Carbon dioxide or Carbon dioxide and oxygen

[a]In hospitals, cyclopropane is often found in chrome-plated cylinders with orange labels or tags.

TABLE 5–3.   DOT Green Label

Gas	Flammable?	Toxic?	Support combustion?
Chlorine	No	Yes	Yes
Oxygen	No	No	Yes
Sulfur dioxide	No	Yes	No
Ammonia	Yes	Yes	No
Carbon dioxide	No	Somewhat	Rarely
Helium	No	No	No

system is a bone of contention between the DOT and the NFPA (National Fire Protection Association), Standard 704, which prefers its own color coding system for hazardous materials. Indeed, it seems a green DOT label can cover a multitude of hazards .The presence of a green DOT label is of little help in determining the hazards of an unknown, escaping gas (see Table 5–3).

A red DOT label at least indicates that the gas is flammable, although, once again, *no differentiation is made between toxic and nontoxic gases.* However, if the material in a package has more than one hazard classification, the package must be labeled for each hazard. A poisonous gas must include a poison gas label. This is in accordance with the DOT general guidelines on the use of labels, Ref. Title 49, CFR, Sec. 173.403(a). Fluorine, hydrogen, hydrogen sulfide, acetylene, cyclopropane, ethylene, methane, and other gases are all shipped under the red DOT label. Their properties and reactions in a fire are as different from one another as those with a green label. Hydrogen cyanide and some other gases are shipped under a DOT poison label.

Some gases have immediately identifiable odors. Ammonia is unmistakable. So are chlorine, sulfur dioxide, and the garliclike reek of acetylene. Other odors can be faint, unfamiliar, or masked by the smell of smoke. The bitter almondlike odor of hydrogen cyanide may kill you if you try to inhale enough to identify it. Hydrogen sulfide, H_2S, has a rotten egg odor, but if a person is exposed to lethal concentrations of the gas, the olfactory nerves are immediately paralyzed and the gas will appear to have no odor at all. All in all, identification of gases by odor is of very limited value.

Hazardous Properties

A gas can support combustion; be flammable, unstable, explosive, corrosive, or toxic; or combine some or all of these unpleasant traits. Although acetylene and natural gas (methane) are always in the fire-cause top ten, gases generally cause fewer fires than flammable liquids. The primary reason is that they are less common. If anything, flammable gases as a group are more dangerous than liquids, not only because of their properties but also because firefighters are not as familiar with the way they will act.

Flammable gases have no flash point. They are ready to burn without preheating of any kind. Their flammable limits are generally wider than those of flammable liquids. Table 5–4 (p. 89) compares the flammable ranges of five hydrocarbon gases, liquified gases, and liquids.

Notice that the liquids have lower explosive limits than the gases but become "too rich" much faster. Only a few liquids can compare with the flammable range of acetylene. Gases differ from flammable liquid vapors in another important respect: All vapors are heavier than air. Some gases are lighter; some heavier. This means that explosive concentrations of

PHOTO 5-4.   DOT flammable gas labels and placards (cont.) DOT label specifications indicate that each diamond (square-on-point) label prescribed must be at least 4 inches (101 mm) on each side with each side having a black solid line border 1/4 inch (6.3 mm) from the edge. The flammable gas label shown above has a red background with the printing and symbol in black.

PHOTO 5–5.   DOT flammable gas labels and placards (cont.). The new flammable gas placard, required when a shipment has a gross weight of 1000 pounds (454 kg), must be red with white symbol, inscription, and 1/2-inch (12.7-mm) border.

88   FLAMMABLE HAZARDOUS MATERIAL

TABLE 5–4. Flammable Ranges of Typical Hydrocarbon Gases and Liquids

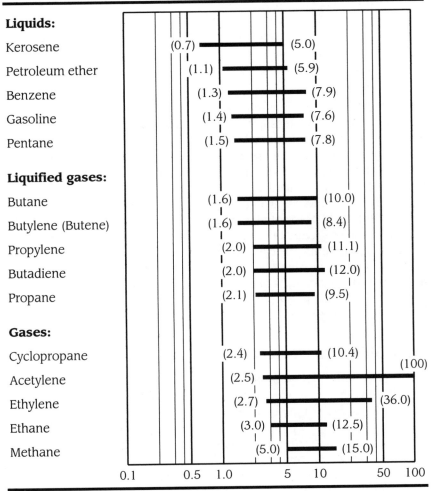

Liquids:	
Kerosene	(0.7) —— (5.0)
Petroleum ether	(1.1) —— (5.9)
Benzene	(1.3) —— (7.9)
Gasoline	(1.4) —— (7.6)
Pentane	(1.5) —— (7.8)
Liquified gases:	
Butane	(1.6) —— (10.0)
Butylene (Butene)	(1.6) —— (8.4)
Propylene	(2.0) —— (11.1)
Butadiene	(2.0) —— (12.0)
Propane	(2.1) —— (9.5)
Gases:	
Cyclopropane	(2.4) —— (10.4) (100)
Acetylene	(2.5) ——
Ethylene	(2.7) —— (36.0)
Ethane	(3.0) —— (12.5)
Methane	(5.0) —— (15.0)

0.1   0.5 1.0   5   10   50 100

certain gases (hydrogen is the perfect example) can collect in the elevated portions of confined areas. Restriction of ignition sources and venting requirements are entirely different for gases and liquids. Table 5–1 on pages 84–85 gives the fire characteristics of six flammable gases.

Pressurized Gas Cylinders

In the introduction of this chapter, the connection between temperature and pressure and how quickly a fire can raise the interior pressure of a gas cylinder to dangerous levels was explored. The way gas pressures are increased was discussed. Firefighters are often called upon to reverse this process, to reduce pressures to safe levels. This can be accomplished in three ways.

1. Increase the volume of the container. Because the only way a steel container can increase its volume is by bursting, this is the least desirable way to reduce pressure.
2. Lower the temperature and pressure of an exposed cylinder by applying a cooling water spray or physically removing the container from a threatened area.
3. Reduce the amount of gas within a cylinder with pressure-relief devices.

Most cylinder valve assemblies have frangible (breakable) discs, designed to break at pressures below the bursting point of the cylinder. While the escape of flammable or toxic gases in a fire situation is a gloomy prospect, it is the lesser of two evils. Poisonous gases may or may not have these relief devices. Chlorine and sulfur dioxide cylinders generally have them, ammonia sometimes does, while hydrogen cyanide does not. Acetylene cylinders have a unique system of pressure relief.

Another hazard associated with high-pressure cylinders is "rocketing." This occurs after accidental rupture or when a value assembly breaks off. Driven by its contents, the cylinder becomes a missile. Cylinders take a little time to build up speed but, once launched, have been known to go through automobiles and concrete walls. The recommendation in emergencies of this type is *get out of the way*. Table 5-1 (page 84) also gives typical cylinder pressures.

Controlling Gas Emergencies

Fortified with knowledge of the type of gas, its hazards, and the dangers associated with the cylinders that contain it, firefighters have the ability to deal intelligently with the emergency.

First of all, protect people. If a toxic or unignited flammable gas is escaping, station firefighters to the windward side if possible, and evacuate everyone downwind. Depending on atmospheric conditions, a single cylinder can make a wide area unhealthy, especially if the gas is heavier than air. This is a good time to remember that water-soluble gases can be diluted and many other gases pushed around with fog patterns.

It will also be helpful to remember which gases are lighter than air (vapor density of less than 1.0). Memorization of the mnemonic "HA, HA, MICE!" will help.

H, helium

A, acetylene

H, hydrogen

A, ammonia

M, methane

I, illuminating gas (a gas mixture of ethane and methane)

C, carbon monoxide

E, ethylene

Not included in the list is nitrogen. All other gases are heavier than air.

Removing Fuel. Whether the flammable material is a solid, a liquid, or a gas, one of the basic principles of the fire tetrahedron is the removal of fuel. When you cool a solid below its ignition temperature, you are not only removing heat but also stopping the production of flammable vapors. This is also true when you cool a flammable liquid below its flash point. But flammable gases have no flash point and require no preheating. Shutting off their flow generally requires some physical action on the part of the firefighter: closing a valve, driving a plug, patching a leak. This should be the main objective. If you extinguish a gas fire without stopping the flow, you convert a bad situation into one that is worse. The escaping gas will quickly form an explosive mixture with air and go in search of an ignition source. Almost certainly, it will find one.

Keeping the Gas Confined. Shutting off the flow is the only sensible way to handle the situation. Sometimes it may be necessary to gain control of the flames so that a well-protected firefighter can gain access to valves that shut off the gas supply. Make sure that the proper valves have been identified. It would be preferable by far to allow the fire to continue and burn out rather than increase the flow or start another by turning the wrong valves. Keep the situation cool by using fog streams to absorb heat and lessen the danger to nearby cylinders, piping, and assorted combustibles.

If extinguishment *must* be accomplished, CO_2, dry chemical, or water spray can all be effective in particular cases with particular gases. A pressurized gas can blow through a foam blanket or dislodge an attempt at a covering CO_2 cloud. However, some gas fires cannot be extinguished by any known method.

NATURAL GAS (METHANE)

Discussion of the paraffin series in Chapter 3 centered on the larger molecules from pentane upwards, looking at the hazards of gasoline, kerosene, naphtha, and other common petroleum fractions. The first four members of this series, however, are flammable gases. **Methane** and **ethane** will be discussed in this chapter, **propane** and **butane** in the next.

General Properties of Methane

Methane is potentially explosive and causes many fires. But its toxic hazards may be overestimated almost as much as the explosive hazards of gasoline are underestimated. Murder accomplished by turning on a modern gas fixture is pure fiction. Methane, although not toxic, can kill by suffocation because it displaces atmospheric oxygen. To be a threat to life, methane must reach a 15 percent concentration in air. Any really toxic gas, such as carbon monoxide, formed in large quantities when methane burns is many times more deadly. It would be more efficient if a fictional murderer allowed the gas fixture to burn and closed the vent. A plastic bag over the victim's head would exclude oxygen much better than methane from an oven.

Perhaps methane gets part of its bad reputation from its various alternate names. The dreaded "fire damp" of miners is caused by pockets of methane trapped in the coal beds until set free by mining operations. In addition, methane is often produced by decaying vegetation, bubbling up in slimy swamps and such, where it is called "marsh gas." Scientists estimate that the intestinal venting of animals adds 45 million tons of methane to the air each year.

If methane were used by industry only in its pure form, it would be of relatively minor importance, just another flammable gas with various applications. Methane, however, is the principal component of natural gas. A portion is stored under pressure or liquified (see Chapter 6); the biggest part, however, is used to supply the daily demand of millions of homeowners and industrial users. The hazards of methane have become a matter of daily importance to every firefighter.

In 1963, the Pacific Coast Gas Association published a pamphlet, entitled "Emergency Control of Natural Gas," directed to the fire service. This section on methane will draw heavily from this interesting and authoritative source. Additional comments, when they are necessary, appear in brackets [ ].

Natural gas is only 65 percent as heavy as air. This means that it will rise and diffuse rapidly when it escapes in an open area. When confined in a closed room, the gas will rise to ceiling level. The air in the room will be displaced from the top *downward*. Remember this when ventilating a room: Open the windows from the top.

[Whether pure methane or natural gas is piped into the home, the hazards are similar. The limits of flammability for natural gas are 4 percent and 14 percent. The ignition point is quite high, about 1,100 to 1,200°F (593.3 to 648.9°C). This temperature is reached by pilotlights, flint sparks, matches, or sparks from electrical switches. One homeowner, returning home after a week's vacation, opened his front door, switched on the lights, and was blown from the porch onto the front lawn when his entire house exploded. The house had filled with gas as the result of a slow leak, and the light switch spark served as the point of ignition. For this reason, some states require mercury switches in all buildings of public assembly.]

Notice that these figures for vapor density, flammable limits, and ignition temperature differ slightly from those for methane. Natural gas also contains trace quantities of ethane, propane, butane, and carbon dioxide. Amounts vary with the locations of the gas field. Their presence changes the properties of natural gas.

Burning natural gas produces little smoke, but it does produce a very high radiant heat. Combustibles must be wetted down with a spray to prevent their ignition by this radiant heat. [Natural gas generally burns blue with an occasional orange flash. A yellow flame may mean that soot and excess carbon monoxide are being formed because of faulty combustion. In a wall furnace, this is almost always caused by lint and dust, clogging the burners.]

Natural gas is odorless in its natural state; therefore, gas leaks in the area where it is produced or at pumping or storage stations may have no odor. Usually, the odorant that lets us detect the presence of 1 percent in a given volume of air is added at the distribution point. One percent is far enough below the lower flammable limit of 4 percent to provide a margin of safety in case of a leak. [A good odorant is nontoxic, noncorrosive, not soluble in water, and chemically inert.]

Emergency Encounters with Gas

Gas Escaping Outside. If gas is escaping from the ground, an excavation, an open pipe, a manhole, a sewer, or a vault, clear a safe area around the location and barricade or rope it off. Extinguish all open flames, prohibit smoking, and make certain that electrical switches or similar possible ignition sources are not operated. Check surrounding buildings, including basements in particular, for any presence of gas odors. It may be necessary to restrict or reroute all traffic until gas company workers can bring the gas flow under control.

When gas is escaping from a broken pipe that is accessible, a wooden plug can be used to stop the leak as a temporary measure. Plugs made from redwood and pine are used because they are soft and will conform to an irregular shape caused by a pipe that is not cleanly broken. They should be driven into place by a rubber mallet. Once a firefighter uses a wooden plug to make an emergency stopoff, the gas worker should be notified so that a safe, permanent repair can be made.

The gas company can help you by supplying specific information, identifying the escaping gas and its combustibility and tracing its source. If it is necessary to enter a manhole or vault, check to see if a concentration of combustible gas is present. If not, enter only if standby assistance is available, and keep in mind that the atmosphere may be oxygen deficient. Manholes or vaults can usually be vented by the temporary removal of their covers. [If possible, gas company personnel and firefighters should stand upwind of a leak.]

Gas Burning Outside.   If gas is burning outside, the firefighter should make no attempt to extinguish the fire. Burning gas will not explode. Clear the danger area and barricade or rope it off. *Never operate gas valves in the street in this type of emergency.* Turning the wrong valve can create an emergency in another area worse than the one at hand and further endanger life or property. Spray water mist on any surrounding combustibles if they are in danger of igniting. Do not use water on burning gas at its point of escape. If this point is in an excavation, the hole becomes filled with mud and water, which makes the repair slower and more hazardous.

Whenever or wherever gas is involved, immediately call the gas company. Their personnel have been instructed to report their presence to the fire officer in charge upon arrival.

Gas Escaping Inside.   If gas is escaping inside a building, ventilate the area, starting where the gas concentration is strongest. If gas is escaping in quantity, clear the building of its occupants. Shut off open flame devices by operating manual controls, but *do not* operate electrical switches. [This includes the main electrical switch. Opening this switch to cut off electrical sources of ignition causes a tiny spark at each switch inside the building. This has been the cause of many explosions.]

In some cases, the fire officer in charge may determine that it is necessary to shut off the gas to the building at the service valve. Once a valve is turned off, leave it off and notify the gas company, who should decide the proper time to turn it on again. [Locating a gas leak either indoors or outdoors can be troublesome. Sometimes, leakage of gas into the ground causes visible damage to vegetation. The leak can be located by noticing dry, discolored plants.]

Gas Burning Inside.   If gas is burning inside a building, shut it off at the meter or outside at the curb valve. If the gas supply cannot be safely shut off, keep the surrounding combustibles wet with spray streams until the gas company emergency crews can control the flowing gas. If it appears that inside piping or installations are threatened by a fire inside the building, whatever the origin, the fire officer in charge should determine if it is necessary to turn off the gas.

[As a matter of practice, many fire departments turn off the gas in a burning building until it can be determined that no hazard of gas leakage exists.]

Case Histories

While emergencies involving natural gas are not rare, the wonder is not how *often* it is involved, but how *seldom*, given the amount of natural gas consumed annually. However, when an incident involving natural gas does occur, it is usually very dramatic and has catastrophic results. The following case studies are indicative of the natural gas case emergencies that have taken place over the past three years.

In Edison, New Jersey, on March 24, 1994 at 12:04 A.M. a 36" pipeline carrying natural gas from Maine to Texas ruptured. Ignition of the escaping gas created a 60' crater and sent flames 500' into the air. One emergency scene worker described the situation as "like Times Square on New Year's Eve." Rocks and searing heat caused mass self-evacuation of the sixty-three building apartment complex immediately adjacent to the rupture. Residents fleeing the complex made access to the fire by emergency vehicles extremely difficult. Four apartment buildings were on fire upon arrival of the firefighters. Fire was extending to four adjacent buildings. No serious injuries were directly related to the incident.

On February 3, 1994, in Steamboat Springs, Colorado, a mini-shopping mall sustained major damage when leaking natural gas in frozen ground seeped through the frozen dirt and through the floor of a restaurant kitchen.

At 11:09 A.M. on December 3, 1992, in Aurora, Colorado, a natural gas line ruptured and drifted onto an interstate highway. The gas was ignited by the cars on the interstate, immediately engulfing several cars. Several individuals in the cars were trapped and received serious burns.

Finally, the following incident that took place in Alameda, California on April 13, 1994, closely assimilates the incident cited in a previous edition of this book relative to a natural gas incident that occurred in Brighton, New York, on September 21, 1951. Excerpts are taken from a California Public Utilities Commission, Safety Division Investigation, dated June 22, 1994:

At about 9 P.M. on April 13, PG&E (Pacific Gas and Electric) monitors detected an increase in pressure in the gas system serving Alameda. By 11:45 P.M., the monitors notified PG&E emergency response personnel. Shortly after midnight, calls came in to Alameda's 911 emergency dispatch complaining of loud noises from gas appliances, natural gas odor, and fires....

After responding to three calls of a similar nature and at the direction of the Alameda Fire Department, at 0020 hours, 911 dispatch contacted PG&E's dispatch in Oakland (Dispatcher) to ascertain the cause of the problem and an estimated time (ETA) by which PG&E would have the situation under control. The Dispatcher stated that the problem was a "low pressure system and it's overpressuring," and that 'Pressure crews were on the way."

Soon after speaking with the Dispatcher, 911 dispatch informed the AFD field units that PG&E had a "low pressure problem." This misinterpretation of information by 911 dispatch led many members of the AFD to believe that PG&E misunderstood the nature of the problem.

Up to 0032 hours, the AFD was encountering gas odors and flaring or extinguished pilots. After 0032 hours, the calls received by 911 dispatch began to indicate reports of fires and smoke....

After 0105 hours, the number of calls received by 911 dispatch system had increased to the point where AFD units were only able to respond to fire-related calls, calls with non-fire-related emergencies were instructed to simply shut off their gas at the outside valve pending the arrival of AFD units. At 0110, the AFD requested assistance from the Oakland Fire Department and the Alameda Naval Base to handle the increasing number of emergency calls.

The incident occurred in a low-pressure system designed to distribute natural gas at approximately one-quarter to one-third psig. The customers on a low-pressure system have meters and shut-off valves, but no service regulators on the gas lines serving their homes or businesses. In most low-pressure systems, gas is supplied by a high-pressure system operating at pressures up to sixty psig, and the gas enters the low-pressure system through a regulator station which contains a regulator, an overpressure protection device, and related appurtenances. Most modern-day service systems are high-pressure systems that regulate pressure at the service entrance to the residence or business.

A high-pressure failure effects only the customer serviced by that regulator. That was not the case in Alameda. For some reason natural gas, at high pressure, was allowed to enter the low-pressure system. The overpressurization throughout the system effected a significant number of homes and businesses. Be aware, the systems are out there and capable of causing widespread problems in communities where they still exist.

ANESTHETICS

When doctors use an anesthetic, a delicate balance must be maintained between the depth of unconsciousness necessary and the medical effect the anesthetic will have upon the patient. All our vital organs—heart, brain, lungs—are regulated by nerve impulses. There must be enough anesthetic to deaden pain but not so much as to interfere with life functions.

Diethyl ether was the first of the anesthetic gases to gain widespread popularity as an anesthetic. Because of its flammability, however, there has been a continual search for better agents. This search led to other, more acceptable gases. However, the flammability factor still remained. Couple the inherent flammability of the anesthetic gases with the oxygen-rich atmosphere of the surgery room and the environment becomes a prime candidate for explosive results. Devastating results from the careless use of flammable anesthetics caused the medical supply industry to develop agents that were nonflammable. However, because of the toxic properties of the nonflammable agents, cyclopropane or ether is still used on certain types of patients. Local fire departments should be aware of any facilities in their area that still employ the use of flammable anesthetic gases. The following information has been retained for that purpose. However, although not widely used in the medical industry, the flammable anesthetic gases are still in production and have many industrial uses.

Ethylene, one anesthetic in use today, is a colorless, flammable gas with a pleasant, sweet odor, It has an important role in the manufacture of organic compounds, ripens fruit and gives them color, and increases the growth rate of seedlings. Ethylene will be discussed again in Chapter 10 on how plastics are made. Figure 5–3 diagrams its structure; Table 5–1 lists its fire characteristics.

Remember: Ethylene will cause unconsciousness on the fire scene as well as in the operating room. *Note:* Ethylene, in sunlight, is sponta-

FIGURE 5–2. Cyclopropane (C_3H_6).

neously explosive with chlorine gas. Storage of these two gases should be separated.

Cyclopropane is another colorless gas. It is a strong anesthetic. Concentrations exceeding one part in 2,500 in air (400 parts per million) can cause unconsciousness. Cyclopropane is shipped in steel cylinders under fairly low pressure. It is another example of the varied structures that the ever-versatile carbon atom will build: this time, triangles. Figure 5–2 diagrams its structure.

Ethyl chloride is not a hydrocarbon. Each molecule contains a single chlorine atom. Occasionally, ethyl chloride is employed as a general anesthetic by doctors, dentists, and veterinarians. More often, it is used for local anesthesia and is applied as a spray. Because ethyl chloride boils at 54°F (12°C), it turns rapidly to vapor, absorbing enough heat from the human tissue it touches to freeze and deaden the area.

Ethyl chloride is also a refrigerant and a solvent and is used in the manufacture of tetraethyl lead. Concentrations over 4 percent (40,000 parts per million) can be fatally anesthetic. It burns with a green flame, forming highly toxic phosgene gas. The flash point is extremely low: –58°F (–50°C).

Despite its sweet odor, prolonged inhalation of **nitrous oxide** can be fatal. It is used only in operations of short duration. Even then, "laughing gas" can have a peculiar influence on patients, hence its peculiar name. While nitrous oxide is not flammable, it contains a higher percentage of oxygen than air and supports combustion more readily. When flammable anesthetics are mixed with nitrous oxide or oxygen their explosive potentiality is increased.

STORAGE AND HANDLING OF ANESTHETICS

Most anesthetics in use today are flammable gases or volatile liquids used in a highly dangerous form, but their use is unavoidable at the present time. If trouble occurs, there is an immediate hazard to life.

The only possible solution lies in a combination of stringent regulation of storage and handling, the complete elimination of ignition sources, rigid observation of safety rules, and our proficiency in inspection and firefighting.

Flammable anesthetics should be stored together in a cool, well-ventilated place outside the area of use. The location should be clearly designated. The following provisions of the National Electrical Code should be enforced.

1. Any room or space in which flammable anesthetics or volatile flammable disinfecting agents are stored shall be considered a Class 1, Division 1 location throughout.
2. In an anesthetizing location, the entire area shall be considered to be a Class 1, Division 1 location which shall extend upward to a level five feet above the floor.

Briefly, the Code separates hazardous occupancies into three classes.

Class 1 includes flammable gases or vapors.

Class 2 includes combustible or explosive dusts.

Class 3 includes easily ignitable fibers.

Class 1 is further separated into four groups.

Group A: atmospheres containing acetylene.

Group B: atmospheres containing hydrogen, or gas and vapors of equivalent hazards.

Group C: atmospheres containing ethyl ether vapor, ethylene, or cyclopropane.

Group D: atmospheres containing most of the common flammable vapors or gases.

Each of the classes has two divisions. Division 1, in each case, has a higher degree of hazard because there is a greater possibility that a hazardous atmosphere will form for one reason or another (see Table 5–5). Therefore, the requirements for approved electrical equipment are more strict.

TABLE 5–5. Relative Flammability of Anesthetic Mixtures

Mixture	Flammable limits in air (percent by volume)		Flammable limits in nitrous oxide (percent by volume)		Flammable limits in oxygen (percent by volume)	
	Lower	Upper	Lower	Upper	Lower	Upper
Cyclopropane	2.4%	10.4%	1.6%	30.3%	1.8%	60.0%
Ethyl chloride	3.8	15.4	2.1	32.8	4.0	67.2
Ethylene	3.1	32.0	1.9	40.2	2.9	79.9
Ethyl ether	1.9	48.0	1.5	24.2	2.0	82.0
Vinyl ether	1.7	27.0	1.4	24.8	1.8	85.5

Class 1, Division 1 locations must have explosion-proof electrical equipment without exception. The five-foot requirement takes into consideration the vapor density of ethyl ether.

Even with explosion-proof equipment, static electricity remains a possible source of ignition. Although tightly controlled by grounding, humidifying the air, and special clothing and shoes worn by doctors and nurses, the generation of static is still the primary cause of operating room explosions. A complete survey of the problem is contained in NFPA Pamphlet 56, "Standards for the Use of Flammable Anesthetics."

Fires involving ethylene, cyclopropane, and ethyl chloride above 50°F (10°C) are best fought like those involving all other flammable gases: Shut off the flow. None of these gases is highly soluble in water. It would be difficult to wash them from the air. Use spray streams to keep the cylinders cool. If a burning cylinder is mounted on an anesthetic machine, move the cylinder to a safe place. Do not overlook the oxygen bottle that may be alongside. If its pressure-relief device fails, burning rate will be increased suddenly. When possible, try to separate the cylinders. If extinguishment is required, dry chemical or CO_2, supported by spray streams, should be effective.

Remember: These are anesthetics. Wear self-contained breathing apparatus.

Finally, as one veteran anesthetist put it, "The fire after an operating room explosion is generally not too severe. The damage has already been done."

Acetylene

If ever the properties and uses of a compound were dictated by its structure, that compound is acetylene, which has within it two carbon atoms joined by a triple bond. Let us see what this looks like and what it means to a firefighter.

Figure 5-3 shows the structural formulas of three hydrocarbon gases that contain two carbon atoms. One is ethane, a member of the paraffin series. Its carbon atoms are linked by a single bond. (All the other bonds are saturated by hydrogen atoms.) Ethylene doubles this linkage, and acetylene triples it.*

The double bond of ethylene contains a great deal of energy. Under the right combination of pressure and heat it can be broken, and ethylene reformed into new products. A triple bond, such as the one in acetylene, is more than just reactive and energetic. It is unstable and potentially explosive. But acetylene has many unique properties that industry can put to work.

*Both ethylene and acetylene are the first members of two series of unsaturated hydrocarbons: the olefins and the acetylenes. The olefins take the suffix *-ene* and include propylene (propane) and butylene (butene). The acetylene series is not well known. Members have the suffix *-yne*. They include propyne and butyne. Acetylene should be called ethyne, but it is not, has not been, nor will it be.

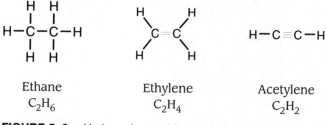

Ethane	Ethylene	Acetylene
C_2H_6	C_2H_4	C_2H_2

FIGURE 5–3. Hydrocarbons with two carbon atoms.

Use and Production of Acetylene

The first thought that comes to mind with the mention of acetylene is of oxy-acetylene welding and cutting. Some large fires have been caused by the bouncing sparks from a cutting torch. Many of these losses could have been avoided by moving the job away from combustibles, by moving the combustibles, or by shielding one from the other. Yet three-quarters of the acetylene produced in this country is used for purposes other than welding or cutting.

Acetylene can be found somewhere in the ancestry of a wide variety of materials; synthetic chemicals, such as acetaldehyde, acetic acid, and acetone; water-based paints; dry-cleaning solvents; and such plastics as Neoprene, Orlon, polyvinyl chloride, Lucite, and Plexiglas. When supplied with the correct amount of air, acetylene burns with a white light instead of its usual smoky flame. This light has many similarities to pure sunlight and can be used in place of electricity, such as in navigation buoys and the carbide lamps of miners.

Acetylene is produced in two ways: by the action of water on calcium carbide, and by the cracking and reformation of other hydrocarbons. Both processes are economically feasible; choice depends on where the plant is located. Where hydroelectric power for electric furnaces is cheap, the carbide process is favored. In oilier portions of the country, acetylene is produced directly from natural gas and propane by "thermal cracking."

When coal and unslacked lime are heated in electric furnaces, calcium carbide and carbon monoxide are formed.

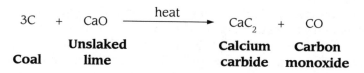

3C	+	CaO	heat →	CaC$_2$	+	CO
Coal		**Unslaked lime**		**Calcium carbide**		**Carbon monoxide**

Acetylene is then generated by the action of water on the carbide.

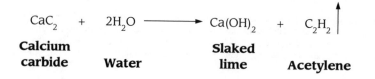

CaC$_2$	+	2H$_2$O	→	Ca(OH)$_2$	+	C$_2$H$_2$ ↑
Calcium carbide		**Water**		**Slaked lime**		**Acetylene**

Inside large metal generators, water and carbide are brought together. Two processes are used: water-to-carbide and carbide-to-water. The latter is preferred because the excess water helps to dissipate the heat caused by the reaction. Acetylene gas is formed, drawn off, purified, and compressed into cylinders. Occasionally, acetylene generators explode from the accidental overpressures, from the ignition of an acetylene–air mixture by glowing carbide or other sources, or from sudden violent decomposition.

Because of the great saving in shipping cost, most commercial manufacturers ship acetylene in the form of the carbide, rather than as gas in cylinders. Each 100 pounds of carbide will generate 31 pounds of acetylene, a pound yielding 4.6 cubic feet of gas. Compare this with five pounds of acetylene in a cylinder with a shipping weight of 100 pounds.

Acetylene Hazards

Pure acetylene is an odorless, colorless gas. Commercial acetylene contains impurities—phosphine, ammonia, and hydrogen sulfide—that give it a garliclike odor. Overall, the toxic hazard of acetylene is relatively slight. While it can have an anesthetic effect in high concentrations, it functions mainly as an asphyxiant.

But there is nothing slight about acetylene's flammable hazard. It has a wider explosive range than any other common flammable gas: 2.5 percent to 81 percent. With oxygen, the upper flammable limit rises to 93 percent. Ignition temperature is remarkably low for a gaseous hydrocarbon, 571°F (299°C). And acetylene burns hot! An oxy-acetylene flame has a temperature of about 5,700°F (3149°C), the highest temperature of any known mixture of combustible gases, a temperature far higher than the flame of an ordinary flammable liquid or solid (See Table 5–6.)

Acetylene is also unstable. If severely shocked or ignited at pressures above 15 psig (pounds per square inch gauge), or subjected to fire temperatures, the triple bond can break apart (decompose), forming

TABLE 5–6. Approximate Flame Temperatures of Certain Flammable Gases

Methane (in air)	3407°F	(1875°C)
Butane (in air	3443°F	(1895°C)
Ethane (in air)	3443°F	(1895°C)
Propane (in air)	3497°F	(1925°C)
Hydrogen (in air)	3718°F	(2048°C)
Carbon monoxide (in air)	3812°F	(2100°C)
Acetylene (in air)	4217°F	(2325°C)
Hydrogen (in oxygen)	4820°F	(2660°C)
Carbon monoxide (in oxygen)	5397°F	(2981°C)
Acetylene (in oxygen)	5710°F	(3154°C)

carbon and hydrogen. Once started, decomposition does not require the presence of air to continue and can quickly escalate into explosive violence. As a consequence, the NFPA forbids the generation and use of free acetylene at pressures higher than 15 psig (a little more than two atmospheres). How then can acetylene cylinders have a standard storage pressure of 250 pounds?

Those who have seen the interior of an acetylene cylinder know the answer: The inside is a calcium-silicate filler made from sand, lime, and asbestos. This filler looks absolutely solid but is 92 percent nothing. It prevents the formation of pockets of free acetylene. In addition, each cylinder is filled with acetone before acetylene is pumped into it. Dissolved in acetone, acetylene is stable. For each atmosphere of pressure, acetone absorbs 25 times its own volume of acetylene. At the standard charging pressure of 250 pounds (about 16 atmospheres), a cylinder contains 425 times its volume of acetylene. In summary, a modern acetylene cylinder contains an almost solid-looking filler, whose microscopic pores are filled with acetylene dissolved in acetone. Were it not for this, acetylene would not be an important industrial gas. It would be too dangerous to use.

Acetylene cylinders have another safeguard. Instead of frangible discs, acetylene cylinders have soft plugs designed to melt around the boiling point of water. Depending on the size of the cylinder, there may be one to four of these soft plugs. All of these specialized safeguards and hazardous properties can cause a variety of fire problems when acetylene is involved.

1. When one of the soft plugs melts, a sudden jet of white flame may extend a dozen feet into the air as acetylene and acetone vapor escape. (This flame will diminish in less than half an hour even though the bottle is full.) If a large number of cylinders are stored together, it can look like Cape Canaveral as all the soft plugs melt. This is the time for discretion, the use of deluge sets for the protection of exposures, and the full support of sprinkler systems.

2. When in use, acetylene is almost always stored next to oxygen. The two cylinders may even be chained to each other. If the acetylene bottle is on fire and the pressure-relief device on the oxygen bottle is still intact, direct every effort toward keeping the setup cool with the use of spray streams. Otherwise, there is the possibility of explosive rupture of the oxygen cylinder or the sudden appearance of a very large oxy-acetylene flame when its frangible disc bursts.

3. Decomposing acetylene can propagate back into its cylinder. The cylinder may overheat to the point where the metal fails. Unless a cooling spray is kept upon it, it will probably explode.

4. When bottles are knocked over, there is also the possibility of a running acetone spill fire if the soft plug melts. This may complicate matters just enough to make the situation impossible. If this happens to a single bottle, the use of dry chemical, supported by fog, can quickly extinguish the acetone fire.

Basic firefighting techniques for a flammable gas still hold true for acetylene, although the problem of how to shut off the flow may get a little complex. *An acetylene fire should not be extinguished except to facilitate the immediate shutting off of the flow of gas.* Be careful about moving a heated cylinder, even if the fuse plug has not melted. If a leak is not on fire, it may be possible to stop the flow in some manner and get the cylinder outside.

The storage area of acetylene must have Class 1, Group A electrical equipment. For obvious reasons, cylinders must be stored upright in a cool, well-ventilated area, preferably protected by a deluge sprinkler system.

Finally, just to prove it is a triple threat, acetylene also forms explosive compounds with copper, silver, mercury, and chlorine. It should never be stored near chlorine, and piping containing acetylene should always be made of iron or steel. The only time acetylene should touch copper is in the head of a torch.

HYDROGEN

There are eleven elemental gases. Seven are considered more or less inert: helium, neon, argon, krypton, xenon, radon, and nitrogen. Three will support combustion: fluorine, chlorine, and oxygen. Only one, hydrogen, is flammable. Indeed, its flammable range of 4 percent to 75 percent is almost as wide as that of acetylene. Hydrogen burns in air with a high flame temperature, 3,700°F (2,037°C); it burns even hotter when part of an oxy–hydrogen mixture. (See Table 5–6.) Most of the energy of burning hydrogen is released as heat rather than light. It burns with a short, intense, pale blue flame that is almost colorless and difficult to see in daylight.

Hydrogen is the lightest element. Although astronomers estimate that hydrogen makes up 90 percent of the universe, free hydrogen gas is relatively rare in the earth's atmosphere. Being only 7 percent as heavy as air, hydrogen "leaks" into space, for the earth does not have enough gravity to hold it. In nature, the hydrogen that remains is locked into such compounds as water, alcohols, and hydrocarbons; hydrogen gas is largely a synthetic product.

Storing hydrogen is difficult. The hydrogen molecule, H_2, formed by two hydrogen atoms, is smaller and lighter than any other single atom. As a consequence, hydrogen tends to leak; it will easily find the tiniest opening in a pipe or container. Pipe threads and valve stem packings must be tight. A high-pressure hydrogen leak, such as from a cylinder at 2,000 psi, can ignite spontaneously. Ignition may be caused by the friction of escaping molecules.

All flammable gas leaks are dangerous but especially so if, like hydrogen, the gas is odorless and colorless. In an enclosed space, hydrogen diffuses quickly and rises to the top. Although explosive mixtures are rapidly formed, if the leak is small, hydrogen may dissipate so rapidly that an explosive concentration may not be reached. But, because it burns with an almost invisible and intensely hot flame and has a wide explosive range that makes for extreme ease of ignition, the hazards of hydrogen gas should never be underestimated. Hydrogen is not toxic and, because asphyxiating concentrations should be found only near the ceiling, its health hazard is not considered severe.

Hydrogen is produced industrially in several ways: by passing steam over hot coal or coke to form **water gas**, a mixture of hydrogen and carbon monoxide; by thermal decomposition of natural gas; by breaking apart the ammonia molecule; or by passage of an electric current through water (electrolysis). The increasing use of liquid hydrogen will be discussed in the next chapter.

Firefighters should know that certain acids, when reacting with metals or when heated, release hydrogen. This can be troublesome if pressure builds up in drums. Hydrogen is also released by the negative elements of lead storage batteries while they are being charged. An investigation by the U.S. Bureau of Mines showed that flammable concentrations of hydrogen existed in a percentage of the battery rooms tested. Many small fires and explosions have occurred when the gas ignited. In some cases, batteries have blown apart, scattering the acid. Adequate ventilation of battery rooms near the ceiling level is important.

Hydrogen is used in welding as part of the oxy-hydrogen flame;* for the hydrogenation of vegetable oils (to turn Wesson Oil into Spry, add a cup of hydrogen); for the synthetic production of ammonia and methanol; and for the cooling of large electrical generators.

Fighting a hydrogen fire can proceed as with other flammable gases. Once again, *shut the flow of gas off.*

SUMMARY

Some pressurized gases are flammable, others are nonflammable. Some support combustion while others are toxic or poisonous. Still others can have a combination of two or more of the aforementioned properties. Add to the inherent hazardous properties of compressed gases the fact that they are packaged in pressurized vessels and the hazards should be apparent. Flammable pressurized gases easily form ignitable mixtures with air and, unlike flammable liquids, tend to move freely, finding ignition sources more readily. They are widely used in industry and in the home, resulting in everyday contact by a wide variety of individuals. Finally, when involved in fire, there is a high degree of possibility that the containers will rupture and become projectiles.

*Flammable mixtures of hydrogen and nitrogen are also sometimes encountered.

REVIEW QUESTIONS

1. Which of the common gases has the widest flammable range?
2. What are the names of four liquids or gases used as anesthetics?
3. What is the primary objective in flammable gas fires?
4. Natural gas consists principally of what hydrocarbons?
5. What gas has an odor of garlic?
6. What are the three ways by which the pressure of a gas is increased?
7. What is the principal cause of operating room explosions?
8. What material reacts with water to form acetylene?
9. What flammable liquid is included inside pressurized acetylene cylinders?
10. What is the lightest gas known to science?
11. Beginning with a gas at normal temperature, how much of an increase in temperature will double its pressure according to the rule of thumb?
12. According to the National Bureau of Standards, an orange cylinder should contain what gas?
13. Name five gases which are shipped under a DOT green label.
14. Name three ways to shut off the flow of gas.
15. What are the toxic hazards of methane?
16. Why is carbon monoxide often formed by wall furnaces?
17. With what gas is ethylene spontaneously explosive?
18. What highly toxic gas does burning ethyl chloride form?
19. The National Electrical Code separates hazardous occupancies into three classes. What does each of these classes cover?
20. Is ethylene water soluble?

DEMONSTRATIONS

1. If available, a cutaway model of an acetylene cylinder is most instructive. It doesn't seem possible that all that acetone and acetylene can find room inside.
2. The differences in vapor density between gases can be demonstrated by two small, compressed cylinders of helium and carbon dioxide and two toy balloons. Fill the balloons. Helium goes up; CO_2 settles to the floor.
3. Generation of acetylene:

   Equipment needed: a deep pyrex or metal dish of fairly small diameter, two candles, water, and some calcium carbide. In the center of the dish, place a heavy holder containing a candle. Fill the dish with water to within a few inches of the candle flame. Drop one or two pieces of carbide into the water. It will fizz, producing acetylene. The gas is immediately ignited by the candle. If the dish is too large, some unignited acetylene may escape. The blackness of the smoke and the heating of the water can be commented upon. The second candle is used to re-light the original candle, which is often extinguished by sudden puffs of acetylene. Practice this first.

BIBLIOGRAPHY

1. *American Heat Video Production,* 240 Sovereign Ct. Ste. C, St. Louis, MO.
   Natural Gas Explosion, Edison, New Jersey, Vol. 8, Program 11.
   Natural Gas Explosion, Steamboat Springs, CO, Vol. 8, Program 11.
   Natural Gas Leak and Explosion, Aurora, CO, Vol. 7, Program 8.
2. National Fire Protection Association, 1 Batterymarch Park, Quincy, MA.
   Standard No. 99, "The Storage of Flammable Anesthetics", 1993 ed.
   Standard No. 704, "Standard System for the Identification of the Hazards of
   Materials", 1990 ed.
3. *Fire Protection Guide on Hazardous Material* (9th ed.), 1986.

VISUAL AIDS

Hazards of Liquefied and Compressed Gases, Big Three Industries, Inc.
Chemical Manufacturers Association, Lending Library, 2501 M St., NW,
Washington, DC.

Hazardous Materials in Your Community; CSX Transportation Company;
Chemical Manufacturers Association, Lending Library, 2501 M St., NW,
Washington, DC.

Understanding Compressed and Liquified Gases. Hazardous Materials
Library, FEMA Region IX; Presidio of San Francisco; FL5-6-393A, VHS
format.

Industrial Gases NFPA. Hazardous Materials Library, FEMA Region IX;
Presidio of San Francisco; FL5-6-367A. (VHS format).

Flammable and Non-Flammable Compressed Gases. Hazardous
Materials Library, FEMA Region IX; Presidio of San Francisco; FL5-4-
186A, VHS format.

Acetylene Cylinders—Handling in Fire Situations. Hazardous Materials
Library, FEMA Region IX; Presidio of San Francisco; FL5-6-416, VHS
format.

Liquefied Gases

When water turns to steam it increases in volume some 1,700 times. While ratios vary to some extent, there is a similar volume relationship between other liquids and gases. A small amount of liquid will produce large quantities of gas. The industrial importance of this fact is evident. By liquefying a gas, we can store large quantities in a small area. This is both convenient and economical.

LIQUEFYING

One method of liquefying a gas is simply to cool it below its boiling point. Cool steam below 212°F (100°C), butane below 31°F (–0.6°C), ammonia below –28°F (–33°C), or propane below –44°F (–42°C), and these gases turn into liquids. The liquid forms because the speed of molecular movement has slowed enough to allow the molecules to adhere to one another.

Critical Temperature

Because it is also possible to force gas molecules closer together by applying pressure, will they adhere if enough pressure is applied? It all depends. As discussed previously, many gases are stored in highly pressurized cylinders. They show no tendency to liquefy. In fact, a gas like hydrogen will not liquefy at room temperature no matter how much pressure is applied. Clearly, there is a factor preventing liquefacation. That factor is called **critical temperature**, the temperature above which it is impossible to liquefy it by pressure alone. The pressure required to lique-fy a gas at its critical temperature is called the **critical pressure.** For instance, the critical temperature of water vapor is 705.2°F (374°C). The critical pressure necessary to liquefy water vapor at this temperature is around 3,200 psi (218 atmospheres). Above this temperature, no amount of pressure will liquefy water vapor. As the temperature drops below crit-ical, so does the amount of pressure required. Finally, at 212°F (100°C), the only pressure needed is supplied by the atmosphere: 14.7 pounds. The reason hydrogen will not liquefy, even under cylinder pressures of 2,000 psi (136 atmospheres), is that its critical temperature is extremely low, close to 400 degrees below zero F (−240°C). In order to be liquefied, hydrogen gas must somehow be reduced to this unearthly temperature. However, liquid hydrogen does exist. How this is accomplished will be discussed in Chapter 7.

However, many gases have critical temperatures well above normal air or room temperatures. The critical temperature of butane is 306°F(152°C), that of propane 206°F (97°C), ammonia 271°F (133°C). If put under suffi-cient pressure in the proper container, these gases will liquefy without cooling. They will boil and produce vapors until enough pressure builds up inside the container to prevent further production; for, as the pressure on a liquid increases, so does its boiling point.* Inside each liquefied-gas container there is a liquid with a pressurized gas above it. As the gas is withdrawn for use, the pressure will temporarily lower. A little more liq-uid boils until the pressure is once again in line with the prevailing tem-perature. In this manner, a gas can be stored at temperatures above its normal boiling point for months or even years. Figure 6–1 compares liq-uefied gas vapor pressure at 70°F (21°C).

Both the dangers and convenience inherent in liquefied gas storage can be readily seen. Containers must be sturdy enough to withstand inte-rior pressures and external shock. If an explosive rupture occurs, all the stored pressure is released at once, and the entire liquid content of the tank boils away. The amount of gas produced will be considerable. In spite of this hazard, the kinds of liquefied gas and amounts in storage are constantly increasing. The most widespread are the liquefied petroleum gases, butane and propane.

---

*Critical temperatures are closely related to boiling points. Butane, with a boiling point of 31°F (−0.6°C) is very close to being a liquid on its own. Just a slight increase in pressure will lique-fy it: 31 pounds at 70°F (1603 torrs at 21°C). But, the relationship between boiling point and critical temperature is subject to some variation: While ammonia and chlorine both boil at −28°F (−33°C), the critical temperature of chlorine is 291°F (143°C), twenty degrees higher than that of ammonia.

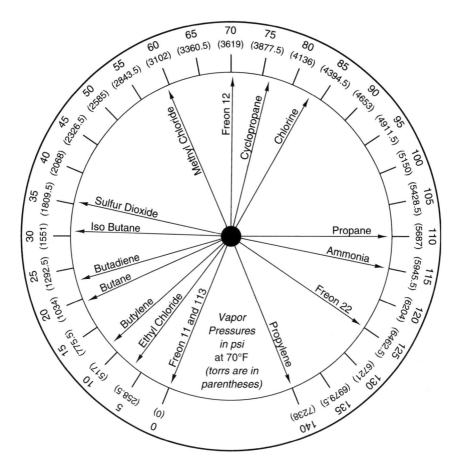

FIGURE 6–1.   Vapor pressures of selected gases at normal temperatures

BUTANE AND PROPANE: THE LP GASES

Liquefied petroleum (LP)* gases have become big business. Annual production exceeds twelve billion gallons. A large percentage of this is used as a fuel gas "beyond the mains," on farms and in house trailers, as a supplement to urban natural gas supplies, and, increasingly, as a motor fuel. Tremendous quantities of butane are used in the production of high-

---

*The technical definition of a *vapor* is a gaseous material that exists *below* the critical temperature, while a *gas* exists only *above* this temperature. Properly, the LP gases should be called the LP vapors in the temperature range between their boiling point and their critical temperature. They never are, which shows how much attention is paid to technical definitions.

octane gasoline. LP gases are obtained as one of the by-products of oil and natural gas wells, or by deliberate formation from other petroleum fractions.

Containers for LP gases range in size from the one-pound, hand-held cylinders often sold in department stores to large metal containers that hold thousands of gallons. A recreational vehicle on the move often carries twin 10-gallon (38-liter) tanks. Another common size is the 28-gallon (106-liter) returnable cylinder with a liquid content of 122 pounds (55 kg). All these containers are built to withstand gas pressures generated at 100°F to 130°F (38°C to 119°C). Most containers also have pressure-relief devices—spring-loaded valves or some sort of fusible plug. Spring-loaded valves are preferred because they will reseat themselves if the overpressure subsides. Once a plug goes it is gone for good. Other methods of protecting LP gas containers include burying, insulation, water-spray protection systems, gas detection and alarm systems, and staggered spacing of groups of tanks.

Propane and butane complete the paraffin series of hydrocarbons. From a firefighter's viewpoint, the worst has been saved for the last. In many ways, they are the most hazardous of the series (See Table 6–1). Although they diffuse fairly rapidly, they are heavier than air and will hang together in a cloud far longer than methane or ethane. At the same time, they are gases under some pressure. As such, they can present a more severe fire problem than liquid hydrocarbons, even such a volatile one as gasoline. The ratio of liquid to gas for propane is 1 to 270. The amount of flammable vapor that propane will produce in a short time far exceeds what would be produced by an equivalent amount of gasoline. In short, the LP gases seem to combine the worst properties of both flammable liquids and gases.

Some other properties of propane and butane require comment. They are almost odorless. For leak detection, a strong-smelling chemical compound called a **mercaptan** is added. Although this smell is distinct, its presence is not always indicative of the presence of LP gas. Other chemicals can emit similar odors. Conversely, chemical reactions require the use of a pure gas, so no odorant is added. It is therefore possible to encounter an odorless LP gas. It should be so marked, but it may not be. Both gases are also colorless. However, a liquid leak vaporizes almost immediately, chilling the air, condensing and making visible the water vapor it contains. Even though the gas is invisible, an LP gas leak can be detected by this vapor cloud. The point of leakage may also be frosted. Sometimes an estimate of the level of liquid in a leaking container can be determined by a ring of frost caused by the rapid vaporization of the liquid as it seeks to restore a pressure balance. When vaporization takes place it absorbs heat from its environment, including the liquid inside the container.

Propane systems depend upon natural vaporization within the container to maintain a steady flow of gas. The amount of heat required for

TABLE 6–1. The Properties of Butane and Propane

	Propane (C$_3$H$_8$)	Butane(C$_4$H$_{10}$)
Flash point (of liquid)	below −100°F (−73°C)	−76°F (−60°C)
Ignition temperature	871°F (466°C)	761°F(405°C)
Flammable limits	2.2% to 9.5%	1.9% to 8.5%
Specific gravity of liquid at 60°F	0.509	0.582
Vapor density	1.6	2.0
Boiling point	−44°F (−42°C)	31°F (−0.6°C)
Water solubility	No	Slight
Volume of gas per pound of liquid (at 60°F and atmospheric pressure)	8.5 cubic feet	6.5 cubic feet
Vapor pressure at 70°F (21°C)	110 psig	19 psig
Vapor pressure at 100°F (38°C)	192 psig	59 psig
Critical temperature	206°F (97°C)	306°F (152°C)
Critical pressure	617 psig	551 psig
Weight per gallon of liquid	4.24 pounds	4.84 pounds
Flame temperature (approximate)	3,497°F (1,925°C)	3,443°F (1,895°C)

vaporization depends upon the rate of use and the climate. If there is heavy usage, the container may require additional heat because the temperature of the liquid may drop close to its boiling point. However, in the United States this is not usually a problem, as propane will vaporize by itself at normal temperatures throughout the country.

Propane has the three qualities needed for a successful LP gas: It is highly flammable, can be liquefied by moderate pressure, and reverts back to a gas at all convenient temperatures. This is not so true of butane.

Butane's boiling point, 31°F (−0.6°C), means that it will not vaporize at many winter temperatures. Therefore, although *butane* has become a synonym for the LP gases, propane and mixtures of propane and butane are much more commonly used, along with such inevitable impurities as isobutane, propylene, and butylene. Propane has the disadvantage of requiring heavier tanks and equipment because of its higher vapor pressure.

Probable LP Gas Emergencies

LP gas emergency situations include a leaking LP container accompanied by a vapor cloud, and an LP container (or containers) on fire, or exposed to fire. Handling of these emergencies requires the following considerations.

1. Protect people. Any vapor cloud will be downwind from the leak. Firefighters should approach from the upwind side, if at all possible. For this reason, places using LP gas should allow access from all sides. The upwind approach is equally important if the LP containers are on fire. Explosions from a newly created vapor cloud can occur. Remove all persons from the area of the cloud or its probable path. Keep back at least 2,000 feet (600 meters) from the area of the cloud wherever it is or goes. The only exceptions to this rule are those people required to deal with the emergency.

Remember: Large LP tanks are *horizontal* tanks. (see Chapter 4, page 68.) Do *not* approach them from the ends.

2. Shut the gas off. This basic rule of gas firefighting applies with even more force to handling LP gas emergencies. There is no tactic more worthwhile than shutting off the flow. Close valves, at the container or remotely, by using valve wheels or wrenches; by crushing or crimping copper tubing; or, as happened on a Hollywood freeway when a tank of butane overturned, by driving redwood plugs into holes. Consult plant personnel or drivers about the location of proper valves. Some containers and systems employ valves that close automatically. If valves cannot be located or used, shut off every ignition source in the path of the vapor cloud.

Remember: An LP gas vapor cloud is *heavier than air* and will sink into low places.

3. Use water to direct the vapor. Although dry chemical will extinguish a very small LP gas fire, *no known method or material will extinguish a large one.* However, the proper use of water will be of assistance. Water is absolutely indispensable. Large amounts should be immediately available. Water can help to protect firefighters closing valves. Without fog patterns, it may be impossible to approach necessary shutoffs. Water can help disperse LP gas vapor. It will not dilute the vapor, but it can push it to a safer location. Fog patterns should be used immediately. Direct the spray across the normal vapor path. If the cloud ignites, there will be a tremendous release of radiant heat that a sufficient amount of fog can help lessen.

Several facts about vapor clouds should be kept in mind: Flames will progress at 15 feet (5 meters) per second through a large cloud, a rate that is about one-half the speed of a desperate person running 100 yards (90 meters) in 10 seconds. If a cloud is seen inside a building, firefighters should not enter except to complete a rescue. An explosion is very likely. A vapor cloud does not necessarily show the limits of the flammable gas, but merely the limits of its refrigeration effect. The flammable gas may extend beyond this on all sides. Therefore, firefighters must keep low behind their fog pattern and should never enter or closely approach the vapor cloud.

If tanks must be removed, protect personnel with water. If a small, leaking portable tank cannot be shut off and must be moved to a safe place, it should be transported in an upright position so that gas, not liquid, will leak. The tank should never be dragged, for doing so can damage valves and piping, possibly increasing the flow. Righting an overturned tank should be done carefully. Above all, keep a spray stream on the tank being moved. Portable containers exposed to heat should be taken to a safe place, but *consider carefully before you move a tank on fire.* It is fairly safe while burning if a cooling stream is kept upon it.

Both the LP gases are nontoxic, but they will cause drowsiness in high concentrations, or produce nausea, headache, or possible asphyxiation. These effects can be avoided by the use of self-contained breathing apparatus.

Water will protect tanks and exposures in the event of a fire. However, the volume of water applied must be commensurate with the magnitude of the problem. A minimum of 500 gallons per minute should be applied to all containers, piping, vessels, exposed tanks, and combustible surfaces. The discharge from burning relief valves can create a giant torch. Care should be taken not to extinguish a vent fire that is not endangering the integrity of the tank. However, when direct flame impinges on the upper area of the tank, the vapor space inside the tank, tank failure is certain if the tank is not cooled by artificial means. This tank failure results in a BLEVE (discussed in Chapter 4). Another benefit of the cooling efforts of master stream appliances is the cooling effect on the LP gas liquid lowering the internal pressure in the tank to a point where vapors cease to emit from the relief value.

The effectiveness of water application will, most likely, be indicated by the sights and sounds of the relief valve. An increase in intensity and noise indicates insufficient cooling of the tank, and immediate withdrawal of all personnel to a minimum of 2500 feet is a must.

Three Case Histories

Three fires involving LP gases are very instructive. Two are indicative of its potential for devastating results and unpredictability. The last is an example of how the study of case histories helps to reduce the possibility of adverse effects' repeating themselves. The case histories discussed in this book are presented in an effort to learn from the tragedies of others. In most cases, individuals thought they were making the right decision. Our responsibility to the memory of those who have died trying to protect life and property is to ensure that their lives and deaths did not go unnoticed or the lessons go unlearned.

On July 5, 1973, the Kingman Fire Department in Kingman, Arizona, responded to a fire in a railroad tank car loaded with liquid propane. The tank was in the process of being offloaded at the time of the incident. Consequently, the liquid line at the protective housing failed, resulting in a propane-fed fire directly impinging on the upper portion of the tank car. (See photo 6–1.) The nearest hydrant was 1,200 feet away. Nineteen minutes after the fire started, the tank car ruptured, and a BLEVE occurred. (See photo 6–2.) The ensuing fireball engulfed three firefighters who were in the process of setting up a deluge gun just fifty feet from the burning tank car. In all, four firefighters were killed instantly, and eight others died days later in the hospital. Only one firefighter in close proximity to the scene survived. As a result of this incident, the Department of Transportation enacted new specifications for tank cars carrying flammable gases. Today, all tank cars transporting flammable gases have thermal protection. However, the possibility of a BLEVE should always be the first issue addressed by the fire officer of any incident involving flammable gases and some flammable liquids.

The next case history is indicative of the powerful, unpredictable and deadly nature of LP gas in ways other than in a fire situation.

On February 24, 1978, some five years after the Kingman incident, there was a train derailment in the small town of Waverly, Tennessee. Twenty-four cars in all were derailed, including two liquid propane tank cars. At the time of the derailment the temperatures were below freezing, and a light snow had been falling. The incident happened at 10:38 P.M. on a Wednesday. There was no immediate release of any toxic or flammable gases or liquids. The next morning, in the light of day, fire department and railroad officials were able to assess the situation. The LPG tank cars were examined, and no leaks were detected. Crews began to move the wreckage, including the two LPG tank cars. They were placed alongside the right-of-way to be offloaded the next

PHOTO 6–1. Vertical flames are caused by the pressure-relief valve venting LPG to the atmosphere. The horizontal flame is caused by a ruptured fill hose. It was the horizontal flame that weakened the tank and caused the BLEVE. (Photo by Hank Graham)

PHOTO 6–2. Kingman BLEVE just after the failure of the tank car. Notice the major portion of the tank car approximately 20 feet in the air. It traveled in excess of 1,000 feet. (Photo by Hank Graham)

day, and the track was rebuilt. By Thursday night the track was reopened and, trains were running once again. On Friday morning, the weather changed and the temperatures reached into the fifties by afternoon. The sun began to beat down on the LPG tank cars, raising the temperature of the liquid gas inside the tank cars. Suddenly, the increased pressure caused a crack in the side of the tank car to open unexpectedly. The escaping liquid gas, turning rapidly to gas vapors, found an ignition source, and five individuals immediately lost their lives. Two died several days later from their injuries. One firefighter underwent numerous operations over the next several years. In this incident, there was no fire, yet the outcome was just as deadly. The critical factor was the movement of tank cars that had possible damage due to the trauma caused by the original derailment.

The third case history is an example of what happens when a fire officer applies his or her education and knowledge to effectively manage a potentially deadly situation.

At 6:30 P.M. on October 1, 1986, the Colorado Springs Fire Department responded to a report of an LPG fire at the local propane distribution plant. On the way into the fire, the engine company witnessed nearly a thousand 100-pound capacity propane cylinders with vent fires fueled by propane gas being released as a result of overpressurization. Almost immediately, the cylinders began to BLEVE. The lot also contained two horizontal cylindrical tanks with 15,000- and 18,000-gallon capacities, each 14 percent full. Throughout the plant property were several more tanks and delivery trucks, all capable of holding anywhere from 250 to 2,600 gallons of liquid propane. The fire ignited as the result of a ruptured discharge hose during the offloading of a 9,600-gallon delivery truck into one of the large storage tanks at the plant. The driver was sprayed with liquid propane, causing frost burns to his face, chest, and right arm. The company officer had to make one very significant and highly stressful decision: fight or flight? This company officer made two very significant assessments. First, he could not evacuate the area quick enough to guarantee safety to the citizens. Second, either the fires were venting vertically or flame impingement was on the bottom of the tanks, in the liquid space. The liquid keeps the tank cool and maintains the integrity of the container. Therefore, he made the decision to set up several large master streams to cool the tanks and avoid BLEVEs. It worked. Twenty-three hours later, the city of Colorado Springs was safe once more. The entire incident is chronicled in the April 1987 edition of *Fire Engineering*.

REFRIGERATION

The principle that causes the air-cooling effect of an expanding LP gas is put to work in refrigeration systems. In the area to be refrigerated, a liquid is allowed to expand into a gas. This expansion takes up heat from the environment. Low-pressure gas, with a high heat content, then leaves the refrigerated space and moves to a compressor. As a high-pressure gas, with a high heat content, it flows from the compressor to a condenser. Here, the heat that keeps it a gas is given off either to the outside air or to a cooling agent of some kind. As it loses heat, the refrigerant condenses into a room-temperature liquefied gas, with a low heat content, that flows back to a receiver and through an expansion valve. Once again, it is allowed to evaporate in the refrigerated space. In this cyclic compression and expansion, the refrigerant transports heat from the area to be cooled to a designed place of discharge. (This is an extremely simplified version of one kind of system. There are many complex variations on this theme.) For a diagram of the system described above, see Figure 6–2. Some common locations of refrigeration systems are:

1. Household refrigerators and freezers.
2. Commercial refrigeration in butcher shops, frozen-food cases, taverns, and restaurants.
3. Air conditioning of all types.
4. Industrial refrigeration in meat-packing plants, produce storage houses, and creameries.
5. Refrigerated transportation in trucks, trains, and ships.

FREONS

From a firefighter's standpoint, the most important variation in these systems lies in the hazards of the refrigerant used. With the exception of ammonia in certain industrial occupancies, and some liquefied nitrogen being used in refrigerated trucks and trailers, practically every refrigeration system being built today uses some form of Freon.

Each refrigerant requires a different amount of compression. Some will produce far lower temperatures in the refrigerated space, and each has a different pressure–temperature relationship. For these and other reasons,

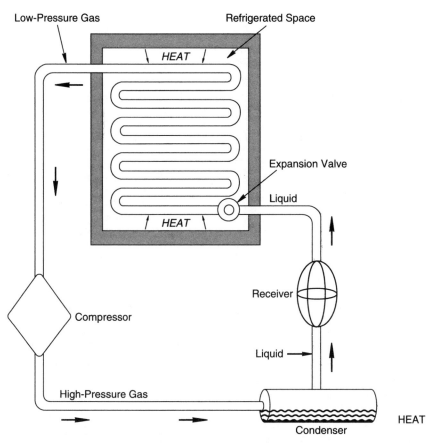

FIGURE 6–2. Refrigeration cycle

there are many kinds of refrigerants instead of only one. Several Freons are currently in use, the most common being Freon–12 (dichloro-difluoromethane.)* A summary of the properties of seven Freons is included in Table 6–2.

A practical knowledge of the types of refrigerants presently being used can be of great value in handling fires, explosions, or gas leaks involving them.

---

*Freon, a brand name of the DuPont Company, refers to halogenated hydrocarbons that contain fluorine. When one or more of the hydrogen atoms in a hydrocarbon is replaced by a halogen (fluorine, chlorine, bromine, or iodine), a halogenated hydrocarbon (halocarbon) results. The jaw-cracking names of the Freons merely indicate what halogens in what number (mono, di, tri, tetra) have hooked on to the carbon atom or atoms.

TABLE 6–2. Summary of Properties of Seven Freons

Refrigerant	Boiling Point	Vapor Density	Temp.-Press. Relationship[a]	UL Toxicity Group[b]	Comments
Freon-11 (CCl_3F)	75°F 24°C	4.7	227 (108)	5	(Trichloromonofluoromethane) Nonflammable
Freon-12 (CCl_2F_2)	−22°F −30°C	4.2	108 (42)	6	(Dichlorodifluoromethane) Nonflammable
Freon-13 ($CClF_3$)	−115°F −82°C	3.6	−2 (−19)	6	(Monochlorotrifluoromethane) Nonflammable
Freon-21 ($CHCl_2F$)	48°F 9°C	3.5	(189) (97)	5	(Dichloromonofluoromethane) Very weakly flammable, ignites at 1,026°F
Freon-22 ($CHClF_2$)	−42°F −41°C	3.0	75 (24)	5	(Monochlorodifluoromethane) Very weakly flammable, ignites at 1,170°F
Freon-113 ($C_2Cl_3F_3$)	118°F 48°C	6.4	280 (138)	4-5	(Trichlorotrifluoroethane) Very weakly flammable, ignites at 1,256°F
Freon-114 ($C_2Cl_2F_4$)	38°F 3°C	5.9	180 (82)	6	(Dichlorotetrafluoroethane) Nonflammable

[a] This is the temperature in degrees Fahrenheit (and Celsius) required for the gas to reach 10 atmospheres pressure, approximately 147 psi.

[b] Underwriters' Laboratories Toxicity Classifications:

1: Concentrations 1/2/% to 1% can cause death or serious injury in 5 min.
2: Concentrations 1/2% to 1% can cause death or serious injury in 30 min.
3: Concentrations of 2% to 2 1/2% can cause death or serious injury in 1 hour.
4: Concentrations of 2% to 2 1/2% can cause death or serious injury in 2 hours.
5: Hazards are intermediate between Group 4 and Group 6.
6: Concentrations of 20% are not injurious after 2 hours.

Explosions can be caused by overpressure of a gas within a system. To prevent these explosions, especially during a fire, refrigerant pressure vessels are equipped with pressure-relief devices large enough to relieve a rise in pressure even though the system is fully involved in a hot fire. Some smaller Freon systems may use fusible plugs that will protect them only against rising temperatures. But there are other causes of overpressures besides temperature rise. An example is the clogging of a frozen part as the compressor forces in more refrigerant. For these reasons, pressure-relief devices are preferred to fusible plugs. All gases are subject to pressure explosions if they are confined and heated, even if, like the Freons, they are almost totally nonflammable.

Toxicity

All the Freons have a low order of toxicity. There is a misconception that the Freons will produce quantities of phosgene gas when exposed to fire temperatures. This probably comes form assuming that the Freons are similar to carbon tetrachloride. Freons do *not* produce significant amounts of phosgene. However, they do produce halogen acid gases like hydrogen chloride. This means that protective equipment and self-contained breathing apparatus are required for firefighters exposed to heated Freons.

When a concentration of 5 percent Freon or more, in air, surrounds large oil or wood fires with no ventilation, it can yield lethal concentrations of acid gas. Acid gases are less dangerous than phosgene because they are very irritating to the nose and throat and therefore easily detected in very small amounts. It is almost impossible to remain, voluntarily, in an atmosphere containing toxic percentages. This is not true with phosgene. Halogen acid gases from burnt Freons can damage metal and clothing. Some situations where this has been observed include the discharge of Freons into the air while a clothes dryer, a stove, or an oven was in operation. In these cases, damage to clothing and metal was evident.

Refrigerant Fires

If refrigeration is involved in a fire, first determine what refrigerant is being used. If it is Freon, pull the main electrical switches to stop the compressors. For air conditioning systems, stop the fans if possible to prevent the spread of fire. *Do not dump the refrigerant.* If the refrigeration piping and vessels are in the fire area, provide ventilation for the burnt Freon acid gases.

On a large system, if Freon is escaping from damaged lines or equipment, try to turn off the main liquid supply valves from the refrigerant receivers. *No harm can come from turning off any or all valves as long as the compressor switch is off.* Maintain a check for leaking refrigerant. As soon as possible, call the refrigeration service company for technical advice. Valves may have to be closed to protect the system against contamination from air or moisture. Perishable products may be in danger of spoilage from rising temperatures. One of the major causes of accidents in refrigeration fires is ice on the floors. Keep water seepage into cold storage rooms to a minimum.

A Freon leak can be detected with the use of a halide lamp. The flame color of the lamp will change to blue or bright green if a small quantity of Freon is present. *Do not use this lamp around flammable gases or other refrigerants.*

AMMONIA

A brief description of ammonia in a chemical dictionary could look like this:

> Anhydrous ammonia, NH_3, a colorless gas with an extremely pungent odor, is highly irritating to the eyes, skin, and respiratory tract. It is flammable, with an ignition temperature of 1204°F (651°C) and a flammable range of 16 to 25 percent. The gas is extremely water soluble and has a vapor density of 0.6. Anhydrous ammonia is shipped in tank cars, tank trucks, and steel cylinders with a DOT green label required. It is widely used in many industrial processes, as a fertilizer and refrigerant.

While these are the necessary bare bones, the hazards of ammonia cannot be appreciated until some flesh is added. The very first word in our brief description requires some explanation: *Anhydrous* means "having no water." At normal temperatures and pressures, anhydrous ammonia is a dry gas with a boiling point of –28°F (–33°C). Because its critical temperature is 271°F (133°C), it can be liquefied by pressure into liquid anhydrous ammonia. Liquid anhydrous is stored like any other liquefied gas in tanks of various sizes, and has its own pressure–temperature relationship. (See Figure 6–3.)

Never confuse liquid anhydrous ammonia with household ammonia, which is something else. When ammonia gas is dissolved in water it becomes ammonium hydroxide or aqua ammonia. While it is true that some ammonium hydroxide solutions, containing 30 percent ammonia gas in water, are much stronger than household ammonia, they are all ammonium hydroxide and less dangerous than liquid anhydrous ammonia.

To summarize, there are three different forms of ammonia. All of them are called ammonia. Do not mistake one for the other.

1. Anhydrous ammonia is the pure dry gas.
2. Liquid anhydrous ammonia is this gas compressed into a liquid.
3. Ammonium hydroxide is gaseous ammonia dissolved in water. (Household ammonia is a weak ammonium hydroxide solution.)

Ammonia Hazards

Anhydrous ammonia is flammable, but its flammable range is extremely high: A 16 percent concentration in air is needed just to reach the lower flammable limit. This is why the DOT classifies ammonia as a nonflammable gas and requires only a green label. This label has caused some

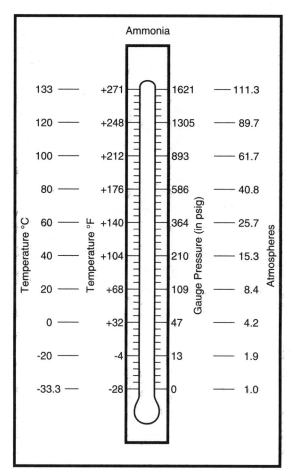

FIGURE 6–3. Temperature–pressure relationship of ammonia

furor because ammonia fires and explosions are not uncommon. Once again, a small amount of liquid will generate a great deal of gas, even enough to reach a lower flammable limit of 16 percent. Yet, even if liquid anhydrous is not present, ammonia gas is explosive. It has even been considered as a possible rocket fuel. The presence of contaminants can greatly increase its hazard.

Nevertheless, anhydrous ammonia doesn't approach the fire hazards of such liquefied gases as propane or butane. Not only is the flammable range higher, but ammonia is lighter than air. Unlike the LP gases, it will diffuse rapidly in an open area. In addition, anhydrous ammonia is extremely water soluble. A sufficient amount of water fog can rapidly turn an explosive concentration of gas into a relatively harmless ammonium hydroxide solution. Ammonia fires are fought like any other flammable gas; the primary object is to halt the gas flow.

In this regard, an ammonia leak or fire can be very difficult to bring under control. It can be a problem to reach and shut off valves because ammonia is far more toxic than the hydrocarbon gases, and personnel must take extra precautions to protect themselves. Fortunately, the piercing odor of ammonia is unmistakable and unforgettable. Anyone not trapped or injured will immediately vacate the area. (This is one case where a wet handkerchief or towel held over the nose can give temporary relief to someone trying to escape.) Low concentrations of ammonia gas can severely injure the respiratory membranes, with fatal results. It causes the throat passages to swell, blocking the airways. The Underwriters' Laboratory (UL) considers it a Group 2 toxic hazard: A concentration of 0.5 percent in air will produce serious injury or death in 30 minutes. All firefighters should be equipped with self-contained breathing apparatus. Canister masks, even those specially designed for ammonia, will filter out only limited percentages.

The liquid–vapor ratio of liquid anhydrous must again be considered, for high percentages of ammonia gas can suddenly appear. Moderate concentrations of ammonia gas can cause firefighters, even if they are equipped with respiratory protection, painful irritation of tender skin at the back of the neck, the forehead above the mask, the insides of the wrists, and the genital area. Gas-tight vapor suits are the only absolute protection, although prompt ventilation and the use of water fog can appreciably lessen the hazard. If desired, water fog will also extinguish ammonia fires. So will dry chemical and CO_2. Liquid anhydrous is even more irritating than the gas and causes severe skin burns on contact. Ammonium hydroxide, especially in the higher strengths, will also cause burns, and its vapors should be avoided by the use of self-contained breathing apparatus.

Ammonia Storage

Liquid anhydrous ammonia is shipped in a wide range of containers under the controversial DOT green label. DOT-approved cylinders are generally required to have a pressure-relief device to prevent overpressure explosions. In the past few chapters, three types of such devices have been discussed.

1. Frangible discs that burst when overpressures occur for any reason.
2. Fusible plugs that melt at designed temperatures, generally somewhere between 157°F and 220°F (70°C and 104°C). These protect the cylinder only against overpressures caused by heat. (Some protective devices combine frangible discs with fusible plugs to prevent bursting at normal temperatures.)
3. Pressure-relief valves that open at a predetermine pressure or heat and close again when this pressure or heat is relieved.

The DOT has several exceptions to its pressure-relief device requirement. Cylinders containing many types of nonliquefied gases stored at less than 300 psi (20 atmospheres) need none. Some, but not all, poisonous gas cylinders need none. Neither do small ammonia cylinders with a capacity of less than 165 pounds. Meanwhile, ammonium hydroxide is shipped in everything from pint bottles to 8,000-gallon (30,000-liter) tank cars. Despite its toxicity, no labels are required, and there are no shipping regulations.

Liquid anhydrous ammonia tanks should be located in cool, well-ventilated places, preferably outdoors. Large containers are equipped with pressure-relief devices. If inside, these tanks should be in a fire-resistive room, underneath a sprinkler system. They should be separated from other chemicals, particularly oxidizing gases, halogens, and acids.

Ammonia is a relatively simple compound of two gases, nitrogen and hydrogen. The molecule is diagrammed in Figure 6–4.

Ammonia decomposes into nitrogen and hydrogen at temperatures between 840°F and 930°F (449°C and 499°C). It is often broken apart, under controlled conditions, to provide supplies of these gases. Both hydrogen and nitrogen have many industrial applications. Nitrogen is used for protective inert atmospheres, in explosives, in many compounds, and, when ammonia is applied directly beneath the ground, as a fertilizer. Ammonia is also valuable on its own, to vulcanize rubber, extract metal from ores, manufacture some types of paper, refine oil, treat water, and form ammonium hydroxide, which is used in pharmaceuticals, soaps, inks, ceramics, and detergents. Anhydrous ammonia is also a very efficient refrigerant.

An ammonia system can be recognized by its steel pipe. Moist ammonia reacts with copper, zinc, tin, silver, and their alloys, causing corrosion. This is why brass or copper piping, soldered joints, or galvanized iron is never used around ammonia. Ammonia compressors are always driven by completely separate motors through v-belts and couplings. The locations of an ammonia leak can be pinpointed through the use of a sulfurous gas that turns white and becomes easily visible.

Fire Histories

There can be explosions of ammonia either at the start of or during a fire. A 16 percent concentration takes some time to build up—enough time for the fire department to arrive at the scene and to begin work. In Iowa, a fire followed a break in a large ammonia pipeline. Probably, the ammonia became mixed with air and lubricating oil vapors. Employees in other buildings saw the white fumes pouring from windows and sent in a fire alarm. The explosion occurred just as two men were entering the engine

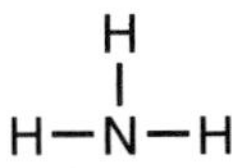

```
    H
    |
H — N — H
```

FIGURE 6–4. Ammonia (NH_3)

room to shut down the generator. Ammonia explosions can be powerful. Many reports speak of windows broken throughout a large area, collapsed roofs, and buildings blown apart.

The explosions can be multiple, each explosion increasing in severity. In a Miami milk-bottling plant, release of ammonia was caused by failure of a 2-inch nipple on an ammonia compressor. The first explosion was relatively minor but attracted the attention of the relief engineer just after 5:00 A.M. His investigation revealed no flame, but a second explosion, ten minutes later, started a raging fire in the engine room. In another ten minutes a third and still heavier explosion spread the fire throughout the building. The loss was $350,000.

Even without an explosion, the ignition of ammonia can accelerate combustion. One report speaks of escaping ammonia gas that resembled a torch as it mushroomed against the ceiling. In many ammonia fires, the intensity became so great it was difficult or impossible for the fire department to get close enough to make an effective attack. Such phrases as "fully involved" or "70 percent of the building was fully involved upon our arrival" are commonplace in histories. Although the upper flammable limit of ammonia is 25 percent, some reports mention the fact that concentrations of gas in the immediate area of the leak were too rich to burn. Fires and explosions were more severe some distance away from the leak where the ammonia was within its flammable range.

The accidental rupturing of an ammonia vessel with the release of unignited fumes is serious enough. Some reports speak of the panic of employees. We can imagine what happened when ammonia fumes became strong within a building where 70 women were employed in processing vegetables.

In spite of our best efforts, tragedies still occur in ammonia fires: At a cold storage plant in Shreveport, Louisiana, on September 17, 1984 firefighters were called to an anhydrous ammonia leak in the warehouse. Emergency response personnel at the scene consulted the Department of Transportation's Emergency Response Guidebook (ERG). At that time, the ERG stated that the most serious hazard was toxicity and corrosiveness. With this information, an assistant chief and captain, along with plant maintenance personnel, entered the warehouse to assess the situation and determine how to shut off the leak. Upon entering the vault area of the leak, the assistant chief noticed a forklift and wheelbarrow containing a white powder. The captain drove the forklift to raise the assistant chief to a height for access to the leaking area. At this point the maintenance men left the area. When fire department personnel noticed that the maintenance personnel had left the area, they exited the building to question the maintenance men about their departure. They said that their suits did not provide enough protection against penetration of the product. The maintenance men assured fire personnel that there would not be a problem with shutting off the leak. The assistant chief felt uneasy but was reassured by maintenance and fire personnel that entry into the building and shutting off the leak was okay. The fire captain and assistant chief re-entered the area and again started to try and shut off the leak. Suddenly the forklift hit something and a resultant fireball engulfed both firefighters in flames. The assistant chief started for a hole blown in the wall by the blast. As he tried to escape through a hole he became trapped. Removing his breathing apparatus and other equipment, he broke free. An all-out effort was made to remove the captain, who was still trapped inside. The assistant chief, burned over 72 percent of his body, survived the ordeal thanks to the efforts of on-scene and hospital personnel. The captain, operator of the forklift, sustained burns over 95 percent of his body and died 36 hours after the incident.

Ignition Sources

Ignition sources for these ammonia fires and explosions vary widely. The list includes oil burners; pilot lights; electric wiring, broken and arcing due to the force of a tank rupture; light bulbs shattered through contact with cold ammonia vapors; static electricity generated by rapidly escaping gas; a system becoming involved in a fire of another origin; and the closing of electrical switches. In Alaska, ammonia filled a building while an employee was checking the oil in one of the compressors. He was forced out of the building by the fumes before he could shut down the compressor. He then went to the transformer building and pulled the main power switch. As he did so, an immediate explosion occurred that raised the roof of the one-story building three feet and involved the entire building in flames.

This history suggests that the pulling of main electrical switches to stop compressors can be dangerous if there is a flammable concentration of ammonia. This is troublesome. One of our primary objectives, if ammonia is escaping, is to try to turn off the liquid ammonia supply valves. No harm can come from closing any or all valves if the compressors are stopped. This is why prompt notification of the refrigeration service company is important. Technical advice may be mandatory. Technicians or engineers of the company in trouble may have been at the scene just before a rupture, a leak, or an explosion. Reports indicate that they are often incapacitated before the fire department arrives, and, to the uninitiated, a refrigeration system can seem a maze of pipes. Sometimes firefighters close the wrong valves during emergencies, thus aggravating the problem.

Ammonia Diffusion Systems

Use of a fire department diffusion system can lessen the danger if an ammonia system is in immediate jeopardy from fire. These diffusers will mix the entire charge of ammonia with water and dump it down the sewer. A fire department box on the outside of the building, preferably in an easily accessible and plainly marked location, contains a valve to dump high-pressure gas and liquid, a valve for the low-pressure gas, and a water valve. It also contains a 1 1/2" or 2 1/2" inlet for additional fire department water supplies. Because a gallon of water absorbs one pound of ammonia gas, the entire amount of ammonia can eventually be mixed into a safer solution of ammonium hydroxide. But this dumping can take some time. Some systems contain more than ten tons of ammonia. *Start the water flowing before the valves are opened to dump the system.* Each fire department should be familiar with the location and operation of all ammonia diffusers within its city. In addition, most ammonia systems have pressure-relief valves located so that they discharge into water tanks or to the air. If they open to air, note their location. Keep personnel upwind and sources of ignition away.

OTHER REFRIGERANTS IN USE

Before we end this section on refrigerants, some others should be mentioned briefly. While they are being replaced by the Freons, some of them can still be found in older systems. Despite their flammable hazards, a few hydrocarbons are used: butane (No. 600), ethylene (No. 1150), propane (No. 290), ethane (No. 170), and isobutane (No. 601).[*] Isobutane has properties similar to those of butane, an ignition temperature of 864°F (462°C), and a flammable range of 1.8 percent to 8.4 percent. Several chlorinated hydrocarbons with varying degrees of flammability and toxicity are still in service. They include dichloroethylene (No. 1130), dichloromethane (No. 30), methyl chloride (No. 40), and ethyl chloride (No. 160).

Sulfur dioxide is considered the most toxic of the refrigerants, the only common one with a UL toxicity classification of 1. In many large cities, it can still be found in the central refrigeration systems of old apartment houses.

Like ammonia, the choking order of sulfur dioxide gives immediate warning of its presence. It is about twice as heavy as air and will linger awhile. It is not flammable. Sulfur dioxide systems may discharge into tanks containing caustic water and lye solutions. Ammonium hydroxide will react with sulfur dioxide to form a harmless white cloud. Household ammonia, in front of blowing smoke ejectors, can sometimes be used when householders become overly enthusiastic in their use of sulfur dioxide fumigants.

Methyl formate (No. 611) is a colorless liquid with an agreeable odor. While it is moderately toxic and irritates the eyes, industrial deaths from this material are extremely rare, occurring only when high concentrations are encountered. Its vital statistics are as shown in the following table.

---

[*]These numbers are the designation given to various refrigerants by the American Society of Refrigeration Engineers (ASRE). They represent the chemical formulas numerically. Ammonia is refrigerant No. 717, and the Freons carry their numbers in the names. Freon-12 is refrigerant No. 12. The usefulness of these numbers is limited. Refrigerants are far more often called by their names.

Flash point	Ignition temperature	Flammable limits	Specific gravity	Vapor density	Boiling point	Water soluble
–2°F	853°F				90°F	
–19°C	456°C	5.9–20%	1.0	2.1	32°C	Yes

Finally, a mixture of two gases, Freon-12 and ethylidine fluoride (respectively, 73.8 percent and 26.2 percent) is now being used as a refrigerant. It is sometimes called Freon-500. It is practically nonflammable, has no warning odor, and has a UL toxicity rating of 5. Once mixed, these gases act like a single substance in that the vapor has the same composition as the liquid. Such a mixture is called **azeotropic**.

SUMMARY

Liquefied gases, unlike compressed gases, are not necessarily hazardous due to their high pressures, but because of their extreme expansion rate and ability to add large quantities of flammable vapors when released. A boiling liquid expanding vapor explosion is graphic evidence of this fact. Most liquefied gases are kept in a liquid form by refrigeration, thereby adding the possibility of cold injuries to the other problems associated with flammable hazardous materials. Although some of the liquefied gases discussed are not highly flammable, their involvement in fire or a leak situation can cause injuries from the toxicity of the product itself or the products of combustion.

REVIEW QUESTIONS

1. What is the difference between liquid anhydrous ammonia and ammonium hydroxide?
2. How may water be used on LP gas fires?
3. What is the difference between critical temperature and critical pressure?
4. When Freons are overheated, what type of gases do they produce?
5. Discuss fire department use of ammonia diffusion system.

BIBLIOGRAPHY

1. Williams, L., and F.A. Carter, "Fight or Flight", *Fire Engineering,* April, 1987, pg.18.
2. Hermann, S.L., "What We Didn't Learn from the Waverly Tragedy", *Fire Engineering,* February 1992, pg.62.
3. Ryczkowski, J.J., "Kingman Revisited", *American Fire Journal,* July: 1993, pg.28.
4. "Dixie Cold Storage Ammonia Explosion", Hazardous Materials Library, FEMA Region IX, Presidio of San Francisco; FL5-6–194A.

VISUAL AIDS

LP-Gas Emergency Planning and Response, National Fire Protection Association; 1 Batterymarch Park, Quincy, MA. (19 minutes, VHS format)

Hazchem 2: Anhydrous Ammonia, Detrick Lawrence Corporation, The Emergency Film Group, 225 Water St., Plymouth, MA. (27 minutes, VHS format)

Propane, Butane, & Propylene, Detrick Lawrence Corporation, The Emergency Film Group, 225 Water St., Plymouth, MA, VHS format.

Cryogenics

Pressurization is not the only way to liquefy a gas. Cooling any gas below its boiling point will also do the job. A gas is a liquid or a solid according to the relationship between its boiling and freezing points and the temperature. We think of water in its three forms only because normal temperatures in temperate zones produce both the solid and liquid states and a hot summer sun evaporates a puddle before our eyes. Yet it is not hard to think of places here on Earth where ice never melts and others where water is vaporized rapidly. The only reason water is predominantly a liquid is that our normal range of climate is neither too warm nor too cold.

Many liquids have boiling points within the temperature range of a hot summer day and a cold winter night. Whether they are gases or liquids depends not only on the month but even on the time of day or night. Methyl bromide boils at 40°F (4°C). It can be liquefied by putting it into a refrigerator. Table 7–1 lists some of these indecisive liquids. Be prepared to deal with them, either as flammable liquids or as gases, with firefighting measures appropriate for each state.

Many gases, particularly elemental gases, have extremely low boiling points. Their critical temperatures are almost as chilly. They cannot be liquefied at ordinary temperatures by any amount of pressure. How can industry have the convenience of liquefied storage at ordinary temperatures when a gas has a boiling point of –423°F (–253°C)? Methods to reduce their temperature to this subantarctic range had to be found, and, once it was found, a technology had to be developed to keep them liquefied although they are stored at temperatures 500 degrees F (278°C) above their boiling points.

The science of **cryogenics** is defined as *the production and use of materials at temperatures ranging from* –150°F (–101°C) *to absolute zero,*

130

TABLE 7–1. Boiling Points of Selected Flammable Liquids

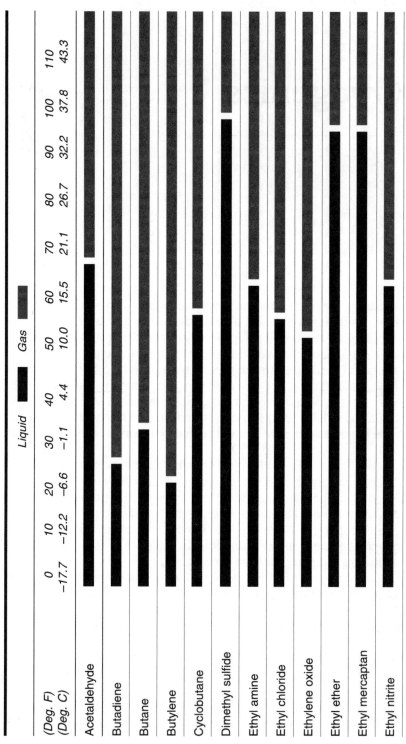

	Liquid ■	Gas ▪

| (Deg. F) | 0 | 10 | 20 | 30 | 40 | 50 | 60 | 70 | 80 | 90 | 100 | 110 |
| (Deg. C) | −17.7 | −12.2 | −6.6 | −1.1 | 4.4 | 10.0 | 15.5 | 21.1 | 26.7 | 32.2 | 37.8 | 43.3 |

| Acetaldehyde |
| Butadiene |
| Butane |
| Butylene |
| Cyclobutane |
| Dimethyl sulfide |
| Ethyl amine |
| Ethyl chloride |
| Ethylene oxide |
| Ethyl ether |
| Ethyl mercaptan |
| Ethyl nitrite |

TABLE 7–1. Continued

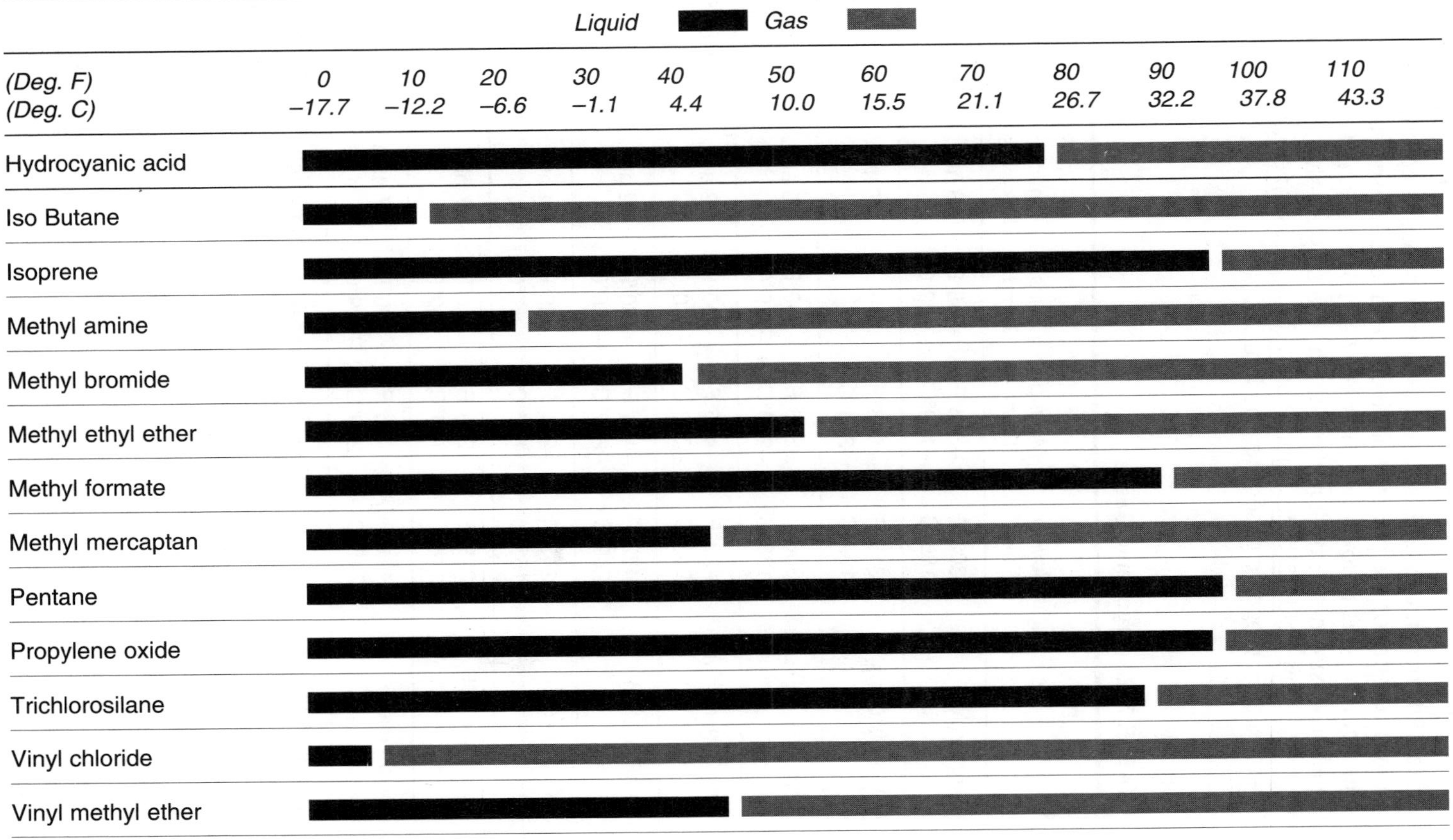

Liquid ■ Gas ▓

(Deg. F)	0	10	20	30	40	50	60	70	80	90	100	110
(Deg. C)	−17.7	−12.2	−6.6	−1.1	4.4	10.0	15.5	21.1	26.7	32.2	37.8	43.3

Hydrocyanic acid

Iso Butane

Isoprene

Methyl amine

Methyl bromide

Methyl ethyl ether

Methyl formate

Methyl mercaptan

Pentane

Propylene oxide

Trichlorosilane

Vinyl chloride

Vinyl methyl ether

131

−459.7°F (−273°C). Producing cryogenic gases, gases at ultralow temperatures, has become an industry with enormous growth potential. One economic reason for the preference given to liquefied gases can be found in the liquid-to-gas ratio shown in Table 7–2. For example, 862 volumes of gaseous oxygen can be liquefied down to a single volume. A cryogenic cylinder can hold 12 times more gas than a pressurized cylinder of the same size. Cryogenic gases are good business.

TABLE 7–2. Cryogenic Gases

	Boiling Point	Melting Point	Critical Temp.	Critical Pressure	Liquid-to-gas Ratio (Volumes)
Argon (Ar)	−302°F −186°C	−308°F −189°C	−188°F −122°C	48 atm.	846 to 1
Air (Mixture)	−318°F −194°C	−363°F −219°C	—	37 atm.	728 to 1
Fluorine (F_2)	−306°F −188°C	−363°F −219°C	−200°F −129°C	55 atm.	981 to 1
Hydrogen (H_2)	−423°F −253°C	−435°F −259°C	−400°F −240°C	13 atm.	840 to 1
Helium (He)	−452°F −269°C	−438°F[a] −272°C	−450°F −268°C	2 atm.	755 to 1
Krypton (Kr)	−243°F −153°C	−251°F −157°C	−88°F −67°C	54 atm.	694 to 1
Methane (CH_4)	−289°F −178°C	−299°F −184°C	−116°F −82°C	45.7 atm.	637 to 1
Neon (Ne)	−411°F −246°C	−416°F −249°C	−380°F −229°C	27 atm.	1445 to 1
Nitrogen (N_2)	−320°F −196°C	−346°F −210°C	−233°F −147°C	33.5 atm.	697 to 1
Oxygen (O_2)	−297°F −183°C	−361°F −218°C	−181°F −118°C	50 atm.	862 to 1
Xenon (Xe)	−163°F −108°C	−169°F −112°C	62°F 17°C	58 atm.	559 to 1

[a] at 26 atmospheres pressure

CRYOGENIC PRODUCTION

The production of liquid nitrogen is a good example of cryogenic production methods. Air is compressed to 1,500 psi (102 atmospheres). As stated earlier, when a gas is compressed, it produces heat. This heat of compression is dissipated in several ways, most cheaply by the use of cold water. The air is now compressed air at 1,500 psi (102 atmospheres) at a temperature of 32°F (0°C). Additional compression raises the pressure to 2,000 psi (136 atmospheres). The small amount of additional heat is again dissipated. The compressed air is then cooled to –40°F (–40°C) by an ammonia refrigeration system. The air is then fed to a larger tank and allowed to expand. The expansion of a gas cools it. As the pressure drops from 2,000 psi to 90 psi (136 atmospheres to 6 atmospheres), so does the temperature, from –40°F to –265°F (–40°C to –165°C).

At this stage, several alternatives are possible: The remaining pressure can be released and the air will drop below its boiling point of –318°F (–194°C). If desired, previously produced liquid nitrogen can be used to cool the air to –320°F (–195°C) before the remaining pressure is released. In this way, the air's temperature can be driven below –400°F (–240°C). Similarly, compressed hydrogen gas, cooled to –400°F (–240°C) before its pressure is released, will liquefy at –423°F (–253°C); helium gas, cooled by liquid hydrogen, can be liquefied at –452°F (–269°C), only eight degrees above the coldest possible temperature.

Once liquid air is formed, it can be allowed to boil off. Because the different gases in air boil at different temperatures, as each forms it can be collected. Or, liquid air can be stored and its various liquid fractions separated according to their specific gravity. For a long time, the most useful and valuable fraction was liquid oxygen. But air is 78 percent nitrogen, which was going to waste. Someone soon found a use for it. In fact, the multiple uses found for liquid nitrogen are a good example of why production of cryogenic gases is constantly expanding.

USES FOR CRYOGENICS

Besides the savings in storage space and weight they offer, cryogenic gases are valuable simply because they are so cold. Liquid nitrogen is used to freeze liquids for emergency pipeline repairs; to solidify gummy materials, such as certain plastics and cosmetics, before they are ground; to preserve human tissue in banks and the spermatozoa of prize bulls before artificial insemination; for shrink fitting of valve inserts, propeller shafts, and bearings; in medicine, for the removal of skin blemishes and the correction of Parkinson's Disease (liquid nitrogen is released deep inside the brain). This listing is incomplete, and other uses, such as the preservation of dead bodies, are found every day.

The use of liquid nitrogen to refrigerate trucks is increasing. A tank inside the storage space expels liquid nitrogen upon command from a thermostat. While nitrogen will combine with several burning metals and is not considered totally inert for that reason, the gas is also used to provide completely safe atmospheres above and around various industrial operations. There would be a small likelihood of any fire inside a trailer when a nitrogen atmosphere is present. If entry into the trailer is required, remember that nitrogen can act as an asphyxiant. Unless you are wearing a self-contained breathing apparatus, open the door and wait three minutes. Look out for metal pipes inside. They will be intensely cold, cold enough to "burn" skin.

Cryogenic gases are not new. The increased use of liquid hydrogen and oxygen in the space program has simply made them more glamorous. Liquid oxygen, nitrogen, and argon have been used, stored, and shipped in large quantities for more than a quarter of a century.

In spite of the fact that firefighters may be unfamiliar with the strange properties and hazards of cryogenic materials, they must learn how to handle them in an emergency.

HAZARDS OF CRYOGENIC GASES

Basically, the hazards of cryogenic gases can be reduced to three categories:

1. *The hazards of the particular gas in question,* although the danger may be accentuated in a cryogenic fluid. Liquid hydrogen and methane are still flammable. Liquid oxygen will support combustion with a roar. Liquid fluorine is reactive and toxic.
2. *The tremendous liquid-to-vapor ratio.*
3. *The extreme cold.*

Because all cryogenic gases have a high vapor ratio and are a freezing threat, these two hazards will be discussed first. Because they are so cold, all cryogenic gases are a danger to flesh: They will "burn" the skin. When spilled on your skin in a small amount, they tend to skitter like a water droplet on a hot stove. In large amounts, they can cling because of their low surface tension. Because these materials are as far removed from your skin temperatures as a hot stove, they can burn you just as badly.

The gases produced by these liquids are also extremely cold. An exposure too brief to injure your hands or skin can damage your eyes by freezing the liquid (largely water) in them. For safety, wear goggles, loose asbestos gloves that can be thrown off in a hurry, and trousers with no cuffs. If any liquefied gas comes into contact with your skin or eyes, immediately flood the area with large quantities of unheated water and apply cold compresses. If the skin is blistered or the eyes are affected, get to a physician immediately.

All of these gases, except oxygen, are asphyxiating or toxic in themselves. There is also the possibility that a gas with a lower boiling point than that of oxygen will extract the oxygen from any air exposed to it, reducing the concentration to levels that will not support life.

In a pamphlet, "Precautions and Safe Practices for Handling Liquefied Atmospheric Gases," Linde Division of Union Carbide Corp., one of the leading producers of cryogenic gases, says:

> Liquid nitrogen is colder than liquid oxygen. Therefore, if it is exposed to the air, oxygen from the air may condense into the liquid nitrogen. If this is allowed to continue for any length of time, the oxygen content of the liquid nitrogen may become appreciable and the liquid will require the same precautions in handling as liquid oxygen. However, most liquid nitrogen containers are entirely closed except for a small neck area and the nitrogen gas issuing from the surface of the liquid forms a barrier which keeps air away from the liquid and prevents contamination.
>
> [But] if enough nitrogen gas evaporates from the liquid in an unventilated space, the percentage of oxygen in the air may become dangerously low. When the oxygen concentration in the air is sufficiently low, a man can become unconscious without sensing any warning symptoms, such as dizziness. If a person seems to become groggy or loses consciousness, get him to a well-ventilated area immediately. If you have any doubt about the amount of oxygen in a room, ventilate the room completely before entering it.

Cryogenic liquids will refrigerate the moisture of the air and create a telltale visible fog except when the air is very dry. Unlike the LP gases, this fog normally extends over an area *larger* than the area containing an appreciable amount of released gas. A hydrogen cloud can be several times larger than an ignitable concentration of gas. Reports show that the percentage of oxygen on the edge of a liquid oxygen fog is very close to that of normal air.

Storage of Cryogenics

Because of cold and the vapor ratio, cryogenic liquids cannot be stored in just any container. Their tanks are double-walled, gigantic vacuum bottles. An insulating material, Perlite, or a multi-layered, crinkled, aluminized Mylar film under high vacuum fills the space between the inner and outer jackets. The inner tanks are built of corrosion-resistant stainless steel, aluminum, or various copper alloys. Carbon steel becomes too brittle at these temperatures. Tanks come in varying shapes—vertical, spherical, and cylindrical—and in sizes ranging up to 175 thousand barrels (28,000 cubic meters).

Heat from the atmosphere will penetrate even a well-insulated tank. The temperature differential may be over 500°F (269°C). Some of the liquid will absorb this heat and vaporize, but most of the fluid remains at the same temperature. Through this process of self-insulation, the temperature inside a cryogenic container remains surprisingly uniform. So does the vapor pressure. During normal operation, the vapor pressure inside a cylinder will not exceed 100 psi (6.8 atmospheres). But if the tank is unused, more and more gas is formed as heat leaks in. The pressure inside the container will rise slowly.

Pressure-relief Systems

1. A **pressure-relief valve** that opens well below any pressure that would threaten the integrity of the tank itself. If the tank is not used for four or five days this valve will open and release the overpressure.
2. A **frangible disc** that ruptures if the pressure risk is excessive or the pressure-relief valve sticks for any reason. This disc is set above the liquid level. It should vent gas only if the container is upright. The frangible disc will quickly lower the interior pressure of the tank to atmospheric levels. It will operate when the insulation is damaged and the inner tank exposed to normal temperatures. In such a case, the evaporation rate may exceed the capacity of the relief valve.
3. The **insulation space** between the inner and outer jackets is protected by a relief device that functions at a few pounds pressure. This guards against leakage from the inner tank. (Lack of such a device may have been the cause of at least one major cryogenic catastrophe.)

In a test, an intact cryogenic container full of liquid nitrogen, protected by these devices, withstood fire test temperatures of 1,800°F (982°C) for an hour. Not even the frangible discs blew, since the relief valves handled the pressure rise. At the end of an hour at 1,800°F (982°C), half the original liquid nitrogen remained, still at –320°F (–196°C). Therefore, if the decision is made to employ a protective water spray to help prevent pressure buildups, care should be taken not to jam the pressure-relief valve with ice.

If cryogenic fluids are trapped anywhere, at any time, they can cause a violent pressure explosion. Given the chance to expand, their vapor ratios make them menacing. This is particularly true in piping. Pressure-relief valves should be installed on every length of pipe between two shutoff valves. All pipes should be sloped up from the tank to avoid the possibility of trapped liquid. Pipes should be made of stainless steel, aluminum, copper, Monel, bronze, or even a plastic that will shatter after a small pressure increase.

CRYOGENICS THAT DON'T SUPPORT COMBUSTION

While all cryogenic vessels exposed to a fire or entangled in an accident should be approached with due caution, several of the elemental cryogenic gases listed in Table 7–2 are similar to nitrogen in that they will neither burn nor support combustion. They are argon, helium, krypton, neon, and xenon. Once called the "noble" gases, these elements used to be considered totally inert and incapable of forming compounds under any circumstances. Recently it was discovered that some of them will combine with fluorine. They have lost their nobility.

Argon is the most common of the inert gases, making up just less than 1 percent of air. Produced in fairly large quantities when air is liquefied, it is used for processing such metals as titanium and zirconium, to fill electron tubes and light bulbs (1.5 billion per year at last count), and to provide inert gas shields.

Helium has several interesting uses besides the filling of dirigibles and balloons. Special helium–oxygen mixtures are provided when individuals must work in high-gas-pressure situations in tunnels or when diving below certain depths. These mixtures help prevent the "bends," caused by the formation of nitrogen bubbles within the body. Helium boils at a lower temperature than any other material. It cannot be solidified at any temperature, even absolute zero, unless pressure is put upon it. More and more liquid helium is being stored and used even though the extremely small difference between its boiling point (–452°F or –269°C) and its critical temperature (–450.2°F or –268°C) makes this difficult. Strange things happen when helium is cooled to temperatures close to absolute zero. There is no friction, no viscosity, no surface tension; there is a boiling without bubbles, and what seems a repeal of the law of gravity. Some metals immersed in it become super-conductors—that is, once started, an electric current never stops.

Neon, krypton, and **xenon** are all rare and quite expensive. Seldom, if ever, will they be encountered in large quantities. When an electric arc is passed through neon it glows red; a mixture of neon, argon, and mercury will glow in various shades of blue and green, while helium glows yellow and white.

LIQUID OXYGEN (LOX)

Even in liquid form, oxygen still will not burn. Somehow this fact never filtered through to the news reporters who wrote the first accounts of the 1967 *Apollo 1* tragedy, complete with details about the "burning oxygen" that enveloped the inside of the space capsule. When more responsible reports were published, it was learned that the materials' burning rates were fatally accelerated by the presence of a pure oxygen atmosphere.

Whenever pure oxygen is present, the sudden ignition and violent burning of combustible materials is possible. The higher the pressure, the greater the hazard. Around pressurized oxygen cylinders, care must be devoted to preventing the sides of the fire tetrahedron from coming together. Such materials as dust, graphite, oil, or grease should never be allowed to contaminate oxygen valves. Cylinders should never be overheated, and valves should be opened slowly to prevent sudden compression (adiabatic heating), which will create heat as surely as the vibration of a loose gasket or the friction of a high-velocity leak.

Whatever can be said about gaseous oxygen supporting combustion goes double in spades for liquid oxygen. While LOX is not flammable either, it enormously increases the burning rate for all combustible materials it contacts. Many substances not normally considered flammable will ignite readily; moderately flammable materials will burn violently; others will explode if mixed with LOX. These potential explosives include all the common hydrocarbons: oil, grease, kerosene, gasoline, acetylene, methane, other flammable gases, tar, dirt contaminated by oil or grease, and even asphalt. There have been reports of liquid oxygen spills that exploded when a fire department hose butt was dropped upon the soaked asphalt of the street. This is why LOX storage should be located on a clean concrete pad.

When such materials as wood, paint, powdered or shredded metals, and cloth are exposed to liquid oxygen, they will also burn violently, or possibly explode, if an ignition source is present. This can happen several minutes after contact. Clothing soaked or splashed with LOX should be removed immediately and aired for at least an hour away from all sources of ignition. Firefighters in uniform or turnout (protective) clothing should be relieved to do this immediately.

No open flames or smoking should be allowed near LOX storage. Burning cigarettes dropped into LOX send up a flame two feet long. In the event of a leak, apparatus and personnel should not be allowed downwind. Pure oxygen can have a destructive effect on a gasoline engine. Everyone not actively engaged in fighting a fire around liquid oxygen should be evacuated. The presence of a fire means that both heat and fuel are present. LOX can escalate the situation into an explosion. Keep an eye on the vapor cloud, for it will not dissipate rapidly until it warms up. Use large quantities of water on fires, on exposed cryogenic cylinders, and to dilute spills. If at all possible, shut off the flow of liquid oxygen.

Liquid oxygen is a pretty, pale blue. It is odorless and tasteless, and its only health hazard lies in its extreme cold. Table 7–2 lists some of its properties. It is the most widely used low-temperature gas. Half the total production is used to manufacture steels by supplying oxygen furnaces or feeding the oxygen lance in open hearth operations. Another quarter produces various compounds for the chemical industry. Sizable amounts are used in certain explosives where organic materials are mixed with it just before detonation. Hospitals use 800-gallon (3-cubic-meter) LOX tanks, with high-pressure cylinders in reserve, in their oxygen systems. Piping carries it throughout the hospital. Only 15 percent of liquid oxygen production is used to oxidize various rocket propellants, but this is how LOX became famous.

CRYOGENIC FLAMMABLES

Two of the gases listed in Table 7–2 are flammable, adding the chance of explosive ignition to the other troublesome properties of cryogenic fluids.

The first of these, liquid hydrogen, is transparent, odorless, and the lightest known liquid, only one-fourteenth as heavy as water. This is one reason why it is produced in large volumes for use as a rocket propellant. By cutting the weight of fuel, it allows rockets to carry more payload. Like other cryogenic fluids, liquid hydrogen is in increasing industrial demand because this is such a convenient way to store the gas. Liquid hydrogen boils at –423°F (–253°C). Only liquid helium is colder. This temperature is so low it cannot only liquefy but also *solidify* oxygen from the air, and solid oxygen can lodge in places where subsequent expansion can cause trouble. If it solidifies inside the liquid hydrogen itself, the combination becomes a high explosive.

Tests conducted by the Arthur Little Company for the U.S. Air Force produced some interesting conclusions about liquid hydrogen. It is not as dangerous to store and handle as the same amount of LP gas. Principally, this is because LP gas gathers in a cloud while hydrogen rapidly rises and dissipates. But, remember that liquid hydrogen, at its boiling point, produces an intensely cold gas that, for a time, is *heavier* than air. As the gas warms, it begins to rise at five feet per second. Quite a large vertical area can be within the flammable range of 4 to 75 percent.

A small leak, however, is likely to dissipate as fast as it is produced. Liquid hydrogen cannot be poured from a small beaker fast enough for it to hit the ground. It flashes into gas almost instantaneously. Thirty-two gallons (121 liters) of liquid hydrogen will evaporate in 30 seconds. In a fire test with 5,000 gallons (19 cubic meters) of spilled liquid hydrogen, flames reached 150 feet (46 meters) into the air. Storage tanks ten times larger than this are increasingly common. Considering its lightness and rate of dissipation, liquid hydrogen should be stored and handled outside whenever possible, away from any overhead enclosures.

Transportation of flammable cryogenic fluids, such as hydrogen and methane, presents the danger of the release of highly flammable gases in the event of an accident. Liquid oxygen, although not a flammable material, requires extra precautionary measures because it so vigorously supports combustion.

Should any of these materials become involved in a motor vehicle accident during shipping, it is important to keep all unauthorized personnel a safe distance away and, if possible, shut off the liquid flow. In the event of exposure to a fire, water should be played on the tank to prevent pressure buildup. Spills should be allowed to vaporize with all ignition sources kept away from the area for at least one hour after the frost has disappeared from the ground. All vehicles transporting cryogenic liquids should be approached with caution if involved in an accident, regardless of whether they carry a nonflammable label or not, especially if the low-temperature liquid is leaking.

Natural gas, liquefied during periods when pipeline transmission is not at capacity, can be stored for vaporization the following winter as a supplemental supply. LNG has impurities that give it a boiling point slightly different from that of methane, which boils at –259°F (–162°C). Several LNG storage tanks contain 175 thousand barrels (28,000 cubic meters) of liquid, the equivalent of 600 *million cubic feet* (17 million cubic meters) of gas.

SUMMARY

Cryogenic fluids can exhibit all of the same characteristics of flammable gases, but then again they may not. Some are flammable, some react violently when they come in contact with an organic material, and some are completely inert. But the two things they all have in common is that they are extremely cold and have an expansion rate far greater than any other form of material. Therefore, in addition to the possible flammability and toxicity problems, cryogenics are extremely cold, and capable of producing large quantities of gas that are either flammable, toxic, or have the potential to form explosive mixtures with almost anything organic. Couple that with the fact that these compounds are under pressure and the potential for disaster is clear.

REVIEW QUESTIONS

1. What are three hazards associated with cryogenic gases?
2. What precautions should be taken before entering a truck using liquid nitrogen refrigeration?
3. What are the health hazards of nonflammable cryogenic gases?
4. What is the most common flammable liquid gas?
5. How does liquid oxygen (LOX) increase the burning rate of materials?

FURTHER READING

1. "Precautions and Safe Practices for Handling Liquid Nitrogen," Union Carbide Corp. (Linde Division).
2. "Cryogenic Storage Vessels," Chicago Bridge and Iron Company.
3. Neary, R.M., "Handling Cryogenic Fluids." NFPA Quarterly.

DEMONSTRATIONS

1. If possible, get some liquid nitrogen. It is readily available, cheap, and the safest cryogenic fluid. *Do not* use liquid oxygen for any demonstration. Formation of the vapor cloud from liquid nitrogen is immediately apparent.
2. Because of the intense cold of cryogenic fluids, hollow rubber balls held in the liquid will pop like light bulbs when thrown against a wall or on the floor. A banana held in the liquid for 30 seconds will drive a nail into *soft* wood. A small amount poured into a beaker will boil off rapidly, leaving a thick frosting on the outside of the glass. By pouring only a few drops into a balloon (use a funnel) you can show the liquid–vapor ratio as the balloon expands and bursts. Practice a little before class; *be sure to wear goggles and gloves.*

VISUAL AIDS

Understanding Cryogenics, Hazardous Materials Library, FEMA Region IX, Presidio of San Francisco, FL5-6-394A, VHS format.

Flammable Solids

Invariably, it is vapors or gases that burn, not liquids or solids. This is why a proper concentration of flammable gas is always ready, willing, and able to explode, and why flash points are so important. Again flash point is the minimum temperature at which a liquid gives off vapor sufficient to form an ignitable mixture with the air near the surface. In other words, it is the temperature at which flammable liquids generate enough vapors to support combustion.

Solids also produce vapors. Some even pass directly from a solid to a gas. This is called **sublimation.** If a solid melts at its ignition temperature, any liquid state that may occur is so brief that it is not worth mentioning. Melting points and boiling points become one and the same.

Ordinary flammable solids such as wood or paper have no liquid states and no flash points. They produce flammable vapors only at their ignition temperatures. These vapors are consumed by fire as rapidly as they are formed. Indeed, without their formation and consumption, no fire is possible. The flow is stopped by "closing the valves," so to speak—by cooling such a solid below its ignition temperature. This not only extinguishes the fire but automatically halts vapor production. Just as the flammability of a liquid is determined by its flash point, and extinguishing methods must be related to this factor, the ignition temperature of ordinary solid materials is the key to their fire control. Most woods, papers, cardboards, and fabrics ignite in the neighborhood of 400 to 800°F (204 to 427°C). Because water is hundreds of degrees below these temperatures, the use of fog, a form of water that absorbs heat very efficiently, will rapidly gain control of fires in most solids, if applied in sufficient amounts in the right places. Sounds easy, but. . . .

The exact ignition temperature of an ordinary solid depends on several variables, one of which is its moisture content. Grass fires are a prime

143

example of this factor. A patch of damp grass, impossible to ignite with a match, burns readily when the whole field goes up. In spite of the dampness, this greater amount of heat vaporizes enough moisture to allow the temperature of the dried grass to be raised to ignition levels. However, in a section of still wetter grass, the fire goes out. It expends too much of its energy evaporating the additional moisture. Ignition temperature is never reached. During a fire, water can be employed to achieve the same result. By wetting and cooling exposed solids, we hold them below their ignition temperatures.

The flammability of ordinary solids also depends upon their shape. Solids burn on their surfaces because these are the only parts that obtain sufficient oxygen. Finely divided solids, having more surfaces available, burn more rapidly and are easier to ignite. A good example of the difference is the way an ordinary campfire is started. Shavings and kindling come first. Shavings can ignite at temperatures 300 degrees F (167°C), lower than those required for a large piece, such as a log, making it possible to get the fire started with a small piece of paper or even one match. A log does not immediately ignite because it conducts heat away from the source so rapidly that ignition temperatures are not reached until the entire large piece is heated. This principle is particularly true for solids that conduct heat readily, such as the flammable metals.

If subjected to temperatures of 350°F (177°C) for 45 minutes, some woods will ignite. They will also ignite in less than two minutes if the temperature is raised to 600°F (316°C). However, there is a lower time limitation. No matter how hot, a heat source must be in contact with a flammable solid long enough to raise it to its ignition temperature or ignition will not occur. This is why people can drop cigarettes into their laps without causing an immediate disaster. Although the cigarette tip is well above the ignition temperature of most fabrics, it is not allowed to remain in contact long enough. When this happens, people move fast. However, when they fall asleep smoking that last cigarette of the day, there is plenty of time.

Wood and paper produce flammable vapors only at their ignition temperatures. The melting point, the boiling point, the flash point, and the ignition temperature of the paper these words are printed on, are one and the same. All are reached simultaneously. But look at the variation in a solid like naphthalene. (See Table 8–1.)

NAPHTHALENE AND CAMPHOR

Naphthalene, either a white crystalline solid or flake that smells like mothballs, was used for this purpose for many years.[*] Moths hate the smell of naphthalene. But so do human beings, so it is being replaced by other smelly materials that moths hate but human beings do not. Even at

---

[*]Although naphthalene has such names as white tar or tar camphor, it will be known as "mothballs" for a long time.

TABLE 8–1. Fire Properties of Camphor and Naphthalene

	Melting point	Boiling point	Flash point	Ignition temp.	Lower flammable limits (% by vol. in air)	Upper flammable limits (% by vol. in air)	Specific gravity (water=1.0)	Vapor density (Air=1.0)	Water soluble
Camphor ($C_{10}H_{16}O$)	354°F 179°C	408°F 209°C	150°F 66°C	871°F 466°C	0.6	3.5	1.0	5.2	No
Naphthalene ($C_{10}H_8$)	176°F 80°C	424°F 218°C	174°F 79°C	979°F 526°C	0.9	5.9	1.1	4.4	No

normal temperatures naphthalene gives off fumes, as those who have ever opened an old trunk will testify. As temperatures rise above normal, so does the amount of vapor in the air. At 174°F (79°C) the percentage is 0.9 percent, naphthalene's lower flammable limit. If an ignition source is present, the time is ripe for it to be effective. If not, naphthalene will melt at 176°F (80°C), two degrees higher, and the liquid will continue to manufacture vapors. At 424 °F (218°C), liquid naphthalene will boil into a gas. This gas, if heat is maintained, will ignite at 979°F (526°C). This is also true for camphor. The major variation is that camphor has a flash point more than 200 degrees F (111°C) lower than its melting point. Camphor melts at 354°F (179°C) but has a flash point of 150°F (66°C).

Remember the ring hydrocarbons, benzene, toluene, and the xylenes? Naphthalene is the big brother of this aromatic series. It is a solid for the same reason the larger molecules of the paraffin series are solids; it packs too much molecular weight to be a liquid. The molecule, a double benzene ring, is diagrammed in Figure 8–1.

Naphthalene is shipped in glass bottles, cans, or burlap bags,[*] or molten in tank barges. It should be stored in a cool place away from ignition sources or oxidizing agents. Naphthalene can be found where dyes, fungicides, explosives, resins, and other materials are being manufactured.

Firefighting techniques include the use of CO_2, dry chemicals, or ordinary foam. Although water can be used to cool the solid or liquid below its flash point, expect extensive frothing and foaming when water comes into contact with hot liquid naphthalene. It may be useful to know that the liquid is heavier than water. Naphthalene vapor is poisonous and irritat-

---

[*]Some materials shipped in solid form can melt at normal or slightly raised temperatures. The problem of containment then arises. A material expected to stay put begins to trickle around. The difference between a liquid and a solid is difficult to pinpoint anyway. What, for example, would a heavy grease or jelly be called? Each will flow in its own good time. These types of materials, often called semisolids, could just as logically be called stiff liquids.

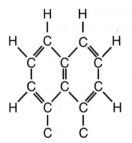

FIGURE 8–1. Naphthalene ($C_{10}H_8$)

ing to the eyes and skin. Avoid contact with the liquid and wear self-contained breathing apparatus and skin protection.

Camphor is a white or colorless crystalline substance with an odor as familiar as that of naphthalene. Commercial camphor is produced by the steam distillation of fifty-year-old camphor laurel trees, native to Formosa but now cultivated in Florida and California. However, most of the camphor used in the United States is synthetic and slightly different from natural camphor; fifty years is a long time to wait. Camphor is used in medicines, in cellulose-nitrate plastics, in insecticides, in tooth powders, and as an embalming fluid. The vapor can be irritating. Camphor is shipped in containers ranging from one-pound (.45-kg) tins to 250-pound (114-kg) barrels or slabs. Storage regulations and firefighting techniques are similar to those applicable to naphthalene.

CARBON

Fires in ordinary flammable solids, in wood and paper and grass, have been and will continue to be the principal reason for the existence of organized fire departments. All the flammable materials examined thus far, the Class "A" solids and Class "B" liquids, woods and hydrocarbons, papers and alcohols, grasses and gases, carbohydrates and ethers, have something in common. They burn, and they contain hydrogen or carbon. It has been estimated that over a million organic compounds (compounds containing carbon) have been identified. Pure elemental carbon takes many forms, some of them extremely useful to American industry.

Coal

The use of **coal** as a fuel and raw material is familiar to everyone. Coal is amorphous (shapeless) elemental carbon.[*] Most coal beds began about 250 million years ago with an ancient forest of trees and ferns. Through

---

*Before you protest that coal has a very definite shape and produce a chunk to prove it, note that "amorphous" describes the internal lineup of atoms. Most solids are crystalline, with their atoms forming definite patterns or lattices, or are amorphous, with atoms hooked up every which way. Although carbon atoms will also crystallize, coal is amorphous (shapeless) carbon.

millions of years they died and were replaced by other vegetation that died in its turn. A mattress of dead vegetation, hundreds of feet thick, built up. The slow geological evolution of the Earth covered this organic matter with blankets of earth and rock. Compression started, and heat increased. Vegetation is principally cellulose, which is about half carbon with varying amounts of hydrogen, oxygen, and trace elements. Gases were slowly forced from the bed, raising the percentage of remaining carbon until lignite, or brown coal, was formed. Lignite still has a high moisture content, and parts of the original plants can sometimes be seen, but its carbon concentration is up to 60 percent.

As pressure and heat were maintained or even increased by further geological changes, bituminous or soft coal was created. There are many grades of bituminous coal, some of which have a carbon content of 86 percent. Many soft coals retain a woody structure. Finally, in beds where compressing pressures were especially intense, hard coals with a carbon percentage as high as 98 percent were formed. The amount of vegetation compressed to form a workable bed of hard anthracite coal, 30 feet (10 meters) thick, is simply stupendous. A one-foot (31-centimeter) layer of coal requires 10,000 years to develop. (The hardness of coal is not necessarily determined by age. Some very ancient beds are still lignite because not enough pressure has been applied over the course of time. **Peat** is carbonized, dead vegetation compressed only by its own weight and water. A peat bog will never make the anthracite grade unless some geological event covers it with a mountain.

Charcoal

If you decide not to wait some 250 million years for a black solid to burn in your fireplace, there is a way to speed up the process: Make a pile of wood, cover it with earth, and set it on fire. In the absence of air, combustion will not be complete. The more volatile gases will be driven off, leaving you with a bed of almost pure carbon, called **charcoal.** Charcoal is made commercially from several carbonaceous (carbon-containing) materials. Animal bones are heated to obtain **bone black**, for example. The charcoal used on weekends to burn in the barbecue is made from various types of wood by modern processes that heat wood to make charcoal, while preserving the valuable gases that boil off. (This same process, using bituminous coal instead of wood, creates coke, a form of carbon that burns with little smoke and such intense heat that it will smelt iron.)

Charcoal has many uses. It will decolorize, deodorize, and filter materials. It is a component of black powder and other explosives. When charcoal is heated to high temperatures, a porous structure with a high capacity for absorbing vapors is created: activated charcoal. Filter-type gas masks contain activated charcoal. The porous structure allows tiny air molecules to pass through but traps some of the larger poisonous molecules.

Carbon Blacks

When hydrocarbons burn, they create quantities of black smoke that are mainly unburned carbon particles. Whenever anything is going to waste, somebody will try to figure out a profitable use for it. **Carbon black** is a powder created from the smoke particles formed when natural gas or liquid hydrocarbons undergo incomplete combustion or thermal decomposition. There are several types.

1. **Channel black** is made by allowing natural gas flames to strike iron plates from which the soot is eventually scraped.
2. **Furnace black** is produced by the incomplete combustion of natural gas or aromatic hydrocarbons in closed furnaces.
3. A similar process, using heavy grade oils, creates **lampblack.**
4. **Acetylene black** results from the incomplete combustion or thermal decomposition of our old friend, the triple threat acetylene.
5. Other hydrocarbons are decomposed to form **thermal black.**

Carbon black strengthens natural or synthetic rubbers and will greatly increase the wear of automobile tires. A typical tire contains five pounds of carbon black and ten pounds of rubber. In fact, anything black probably includes one of these powders, for they pigment inks, dyes, paints, shoe polishes, carbon paper, crayons, typewriter ribbons, and phonograph records. Lampblack is used in insulation systems, explosives, and matches. Acetylene black absorbs liquids more effectively than other carbon blacks and has a higher electrical conductivity. It will convey these qualities into plastics and rubber, or into dry cell batteries. What with one use or another, over two billion pounds of carbon black are used annually in the United States. That's a lot of smoke.

Coal, charcoal, and the carbon blacks are black and amorphous. But carbon can also fashion crystals. If the atoms arrange themselves into a six-sided figure, called a hexagon, the result is a slippery, shiny, black solid called **graphite.** There are large natural deposits of graphite in Sri Lanka and Mexico, and it can be produced synthetically. Powdered graphite is a familiar lubricant. Mixed with clay, graphite is used in "lead" pencils. Pure graphite rods control the reaction in some atomic piles. Graphite powder can be used as an extinguishing agent for certain metal fires because its ignition temperature is quite high: about 1300°F (704°C).

Carbon has another crystalline shape. If subjected to tremendous heat and pressure, carbon atoms will realign themselves into an eight-sided crystal, called an octahedron. This form of carbon is the hardest naturally occurring substance known: a diamond.[*] Although diamonds can be black (a tough, compact, black diamond called carbonado is often used for industrial purposes), they also come in crystals of many colors.

---

[*]When an element has more than one form, the forms are called **allotropes.** Diamond and graphite are allotropic forms of carbon.

Besides gemstones, there are diamond glass cutters, drills, wire dies, and abrasives. A diamond is pure carbon. It will burn. A conflagration in a diamond collection would undoubtedly be the smallest large-loss fire in a department's history.

Hazards of Carbon

Now that the industrially important types of pure carbon have been identified, what is the concern? As always, the hazards of a material arise out of its properties. Carbon atoms "hook" together very efficiently. (See valence, Chapter 2.) Inside a lump of carbon, most of the atoms are firmly attached to four other carbon atoms. The difficulty comes at the edge of the lump where all these "hooks" can be visualized as sticking out, waiting to latch on to something, generally a passing oxygen molecule. When they combine, carbon is oxidized and spontaneously heats. This form of heating is most prevalent in amorphous carbon. (See Figure 8–2.) When carbon atoms are crystallized, fewer possibilities exist in their ordered linked patterns for unsatisfied valences.

With the possible exception of anthracite, all grades of coal are subject to spontaneous heating and ignition. Anything that creates fresh reactive surfaces can cause trouble. As coal is mined, shipped, or stored, it is broken apart accidentally, or intentionally into smaller and more usable pieces. New surfaces are free to oxidize, and more surface is available. Because spontaneous heating is a cumulative process, coal is most dangerous 90 to 120 days after mining. Foreign matter can contribute to this danger in two ways: First, it can produce gases that are picked up by the

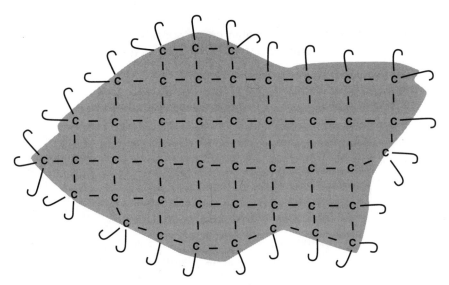

FIGURE 8–2. Schematic of amorphous carbon.

carbon along with oxygen. This combination oxidizes together. Such a mixture may have a lower ignition temperature or may heat more rapidly than pure carbon. Second, heating, an inevitable by-product of oxidation, may cause the ignition of such materials as grease, oil, paper, wood, leaves, or other debris with low ignition temperatures. For these reasons, coal should be stored in a clean location.

As we have learned from the overworked "pile of oily rags" example, spontaneous heating is a cyclic process. More heat means faster oxidation, which, in turn, creates still more heat. Coal should not be stored near heat sources that can give this process a boost. Also, just as a tightly packed pile of rags is less likely to heat spontaneously, tightly piled coal is less likely to heat. What is more important is that alternate wetting and drying of coal aids spontaneous ignition. Coal is a solid that resists penetration by water. Only its face is washed, cleaning off the surfaces and getting them ready for action again. This is also true for charcoal, another potential spontaneous igniter.

Charcoal heats most easily when it is freshly made, ground and exposed to air, or when it is moistened and later dries. Wet charcoal should be stored separately from dry, and inspected frequently. (When wood is exposed to low-temperature heat sources for a long time, charcoal sometimes builds up on its surface. Spontaneous heating of this porous charcoal is suspected as the cause of some mysterious ignitions.)

Carbon blacks are most hazardous immediately following manufacture. This is not only because of oxidation but because bags sometimes contain hot carbon particles. For this reason, carbon black is stored in an observation warehouse before final disposition. After cooling and airing, it does not spontaneously heat. Lampblack, considered the most susceptible to heating, often ignites spontaneously when freshly bagged. Any form of finely divided pure carbon, whether pulverized charcoal, coal dust, or one of the carbon blacks, is a possible dust explosion hazard. Once again, lampblack is particularly hazardous. Coal is sometimes sprayed with a high-flash-point mineral oil that not only reduces dustiness but also protects coal surfaces against oxidation.

Surprisingly, with all the problems it creates, pure carbon is considered a relatively inert element. It ignites in air at temperatures between 600 and 1,400°F (316 and 760°C). When it combines with oxygen, the carbon atom itself does not vaporize. It merely transfers its allegiance to a carbon monoxide or carbon dioxide molecule. To turn carbon into a gas requires temperatures above those produced by normal burning. Carbon starts to sublime (turn directly from a solid to a vapor without bothering about the liquid state) at 6,400°F (3,538°C). At lower temperatures, solid carbon combines directly with atmospheric oxygen to cause a glowing combustion, one of the rare exceptions to the rule that only gases or vapors will burn. With no flame to guide us, the burning process of carbon can be very subtle. Overhaul procedures after a carbon fire must be extremely thorough.

Fighting fires in coal or charcoal can be difficult and dangerous. Water should not be used recklessly. If a small amount of coal or charcoal is involved, it may be possible to separate it and thoroughly wet it down. Otherwise, you may have no other choice but to use quantities of water to reduce the entire pile below its ignition temperature. Remember that pure carbon burns hot. There will be a possibility of a steam explosion within a heated pile. Recognize that all wet material will have an increased ignition hazard when it dries. Advice to the storer or shipper must include consideration of this fact.

When burning, pure carbon produces quantities of carbon monoxide and carbon dioxide. If moisture is present, steam and hydrogen gas can also form. Hydrogen and carbon monoxide are flammable; carbon monoxide is poisonous; carbon dioxide is an asphyxiant. *Pure carbon fires, therefore, are potentially poisonous or explosive, especially if they are indoors* where concentrations of gas can build up. Protect firefighters with self-contained breathing apparatus.

Phosphorus

In the examination of the fire properties of hazardous materials, it is inevitable that at least one would have an extremely low ignition temperature: **phosphorus.** Too reactive to be found free in nature, this element was first discovered by an ambitious German alchemist when he analyzed the solid residue in a sample of urine. Phosphorus has three allotropic forms. Although each is pure phosphorus—nothing added or taken away—they have distinctive appearances and properties. The industrially important allotropes are white (or yellow) phosphorus and red phosphorus (Table 8–2). (Black phosphorus, which looks like graphite, is quite rare.)

White Phosphorus

White phosphorus, when exposed to air in a solid state, ignites at 86°F (30°C); when finely divided, it ignites at room temperature. At normal temperatures, solid white phosphorus combines with atmospheric oxygen and glows. It is "phosphorescent."

How can industry store and handle such a material? Fortunately, phosphorus will not normally ignite when cut off from atmospheric oxygen. It can, therefore, be shipped in hermetically sealed cans, in drums, or in tank cars under either water or an inert gas. Extreme care must be used in handling these containers. Once white phosphorus is exposed to air, ignition is very likely. In storage areas, there should be a periodic check for water leakage. Phosphorus should be stored in a protected area, apart from other materials, particularly oxidizing agents. When mixed with such materials, it can explode.

White phosphorus is a colorless or white waxy solid, often found in sticks about the thickness of a man's thumb. It turns yellow on exposure

TABLE 8–2. Properties and Labeling of White and Red Phosphorus

	Ignition Temperature	Melting Point	Boiling Point	Specific Gravity	Vapor Density	Water Solubility	DOT Label Requirements
White phosphorus	86°F 30°C	112°F 44°C	549°F 287°C	1.8	4.3	None	(Dry) Flammable solid and poison label. Shipment by passenger carrying aircraft or cargo-only aircraft is forbidden. (In water) Same as above except that it may be carried in packages not to exceed 25 pounds (11 kg) per package, on cargo-only aircraft.
Red phosphorus	(over) 393°F 200°C	—	781°F 416°C	2.3	4.3	None	Flammable solid label required. Shipment on passenger carrying aircraft or railcar is forbidden. Packages not to exceed 11 pounds (5 kg) per package may be carried on cargo aircraft.

to light. Solid white phosphorus causes severe flesh burns on contact. Do not let it touch the skin. Inhalation or ingestion of even small amounts (1/300 of an ounce or 850 milligrams) can be fatal. Fumes from burning phosphorus, while not considered highly toxic, are very irritating and extremely dense. Flame-proofed protective clothing and a self-contained breathing apparatus are highly recommended for firefighters fighting a white phosphorus fire. The key to extinguishment is the fact that phosphorus will not burn under water. It is almost twice as heavy as water and not soluble. If drowned with large amounts of water without allowing hose streams to spread the liquid phosphorus, it will solidify below 112°F (44°C) and can be covered with wet sand or dirt. Be careful about the premature removal of this cover. Re-ignitions are very likely. Once the phosphorus has been extinguished, pause and consider before proceeding.

Industrially, white phosphorus is used in certain delayed-action rat poisons, incendiary grenades, and bombs. The use and effects of the 30-pound white phosphorus bomb during World Wall II have been graphically described by Martin Caidin[*]:

> The U.S. Strategic Bombing Survey reports officially that the psychological effect of phosphorus bombs was far greater than any actual damage they caused. British reports of phosphorus bombs dropped by the Germans against English cities indicate that the use of this incendiary material was severe enough in its effect to drive the fire-fighting teams crazy. Whenever it dried out, it immediately exploded into flames. You could cover phosphorus with sand and douse it with water and it would stay quiescent. But the moment the covering was gone, the flames immediately burst out again!
>
> Phosphorus sticks grimly to any surface it touches —metal, wood, concrete, clothing, or human flesh. Whatever the material, the phosphorus will cling to it for good (unless it is scraped off) and will continue to burn itself out unless it is permanently denied access to oxygen.

Caidin's descriptions of human beings caught in a shower of white phosphorus near a river is unforgettable.

> Those with the flaming chemical on their arms and legs and bodies were able to extinguish the flames instantly on entering the water. But those agonized souls who had the phosphorus on their faces and heads! Certainly the flames went out as they plunged into the waters of the Alster. The moment they came up, however, the phosphorus received its oxygen and again burst into flames. And so began the unbelievable terror of choice—death by drowning or by burning.

The following account is from an incident involving elemental phosphorus:

> At approximately 10:50 A.M. on Wednesday, February 20, 1991, 29 cars of a 113-car freight train derailed in Columbia, Missouri. One of four cars carrying elemental phosphorus under water was punctured. The first emergency personnel to arrive on scene witnessed a white cloud rising from a cluster of derailed tank cars. A reconnaissance team in full protective clothing and self-

*Martin Caidin, The Night Hamburg Died (New York: Ballantine Books, 1960).

contained breathing apparatus approached the spill and discovered a two-foot pool of molten phosphorus on fire underneath one of the derailed tank cars. A dense cloud of phosphorus pentoxide gathered in the immediate area, and flames were estimated at two to three inches high. By 3:30 P.M. the fire had become more aggressive, and the cloud was increasing noticeably. The decision was made by the incident commander to attack the fire using foam in an effort to eliminate oxygen from coming in contact with the phosphorus. The foam provided a blanket over the phosphorus and was more effective than just a water application. If only water had been applied, some way of creating a pool effect over the burning phosphorus would have been necessary. The integrity of the foam blanket was maintained by reapplication of foam as it broke down and dissipated. Initial knockdown was achieved with 15 gallons of foam and 250 gallons of water.

Red Phosphorus

Red phosphorus is much less dangerous than white phosphorus. Most important, its ignition temperature is above 392°F (200°C), although this temperature can be easily reached in a fire. Under some fire conditions, it is possible for red phosphorus to revert to white phosphorus. The same care must be exercised against re-ignitions. Again, the use of large amounts of water is recommended. It will produce quantities of dense fumes that should be avoided. Use self-contained breathing apparatus. Red phosphorus does not have to be shipped under water, although both forms carry the same DOT red and white striped label as flammable solids.

Red phosphorus is preferred by industry as a source of pure phosphorus. It is used in the manufacture of phosphoric acid and other phosphorus compounds such as phosphate fertilizers and insecticides. In one form or another, 600 million pounds (273 million kg) of phosphorus are used every year. Most individuals have probably carried phosphorus in their pocket at some time, proving that it does not have the immediate corrosive effect of its allotropic kin. One use of red phosphorus is in the manufacture of matches. It is mixed with glue and ground glass to provide the striker for safety matches. When a safety match is drawn across the surface of this striker, a tiny piece of the red phosphorus ignites by friction and fires the flammable mixture on the tip of the safety match. Ordinary strike-anywhere matches have other phosphorus compounds with a low ignition temperature built into their heads. One such compound is **phosphorus sesquisulfide.**

Phosphorus Compounds

Phosphorus sesquisulfide is the first flammable compound discussed that contains neither hydrogen nor carbon. The molecule, a combination of four phosphorus atoms and three atoms of sulfur, P_4S_3, is a highly flammable mass of yellowish crystals, Its ignition temperature is the same as the boiling point of water, 212°F(100°C). Of course, phosphorus sesquisulfide ignites easily by friction. A thumbnail will light a match, and so will ignition sources not otherwise considered dangerous, such as steam pipes.

PHOTO 8–1. DOT flammable solid label. This red-and-white-striped label is required on containers when flammable solids are shipped. The flammable solid placard for use on motor transports and rail cars is similar in design and color.

Sesquisulfide is not considered a severe toxic itself, but it produces sulfur dioxide when it burns. Use self-contained breathing apparatus. Like pure phosphorus, sesquisulfide is heavier than water and not soluble. Firefighting, once again, can include flooding with water because blanketing is possible. The compound is shipped in glass jars and bottles, cases and steel drums under a DOT red and white striped label and a poison label. It must be shipped separate from flammable gases, liquids, oxidizing materials, or hydrogen peroxide. Storage should be in a cool, protected place.

Phosphorus pentasulfide, P_2S_5, a yellow crystalline substance with an indescribable odor, is used as a lubrication oil, as a rubber additive, and in insecticides. It has several interesting characteristics.

1. Although it has an ignition temperature of 287°F (142°C) and ignites easily, it does not burn rapidly.
2. It reacts with water in several ways. It is decomposed by moist air and can heat or ignite. It reacts with water to form a poisonous and flammable gas, hydrogen sulfide.
3. It is more toxic than sesquisulfide and also forms toxic and irritating gases when it burns.

Therefore, while water can extinguish pentasulfide fires, using CO_2 is preferable. Self-contained breathing apparatus should be worn. Pentasulfide is shipped in glass bottles and sealed drums that should be kept closed whenever possible. It also carries the DOT red and white striped label as a flammable solid.

PHOTO 8–2. DOT poison label. This label must be applied to all packages containing Class B poisonous materials. It has a white background with black lettering and a black skull-and-crossbones symbol. The poison placard that must be shown on motor transports and rail cars is similar in design and color.

SULFUR

Like carbon, hydrogen, and phosphorus, **sulfur** is an element that burns. It is generally a yellow solid, occurring most often as a crystal or powder. If sulfur vapors are suddenly cooled, they form a yellow powder known as flowers of sulfur. Liquid sulfur freezes into solid brimstone. (Brimstone is an ancient name for sulfur.) Despite its bad reputation, pure sulfur is almost odorless. But it is part of many compounds not welcome in polite society. When it burns it forms **sulfur dioxide** gas, which smells as terrible as sulfur is supposed to. Sulfur combines with hydrogen to create **hydrogen sulfide** gas, with an aroma of rotten eggs, or with carbon to make **carbon disulfide,** which smells like a head of wet, year-old cabbage. Hydrogen sulfide (H_2S) carries both the DOT red flammable gas and a poison label. Carbon disulfide carries the DOT red flammable liquid label. The **mercaptans,** a group of sulfur-containing compounds, sometimes used for odorizing natural gas, are also found in the working end of a skunk. (See Table 8–3 for a summary of the properties of sulfur. Several other sulfur compounds are also included or briefly summarized later in this section.)

As discussed earlier, flammable solids are very versatile in finding ways to burn. Naphthalene and camphor produce flammable concentrations or vapor below their melting points. They are solids with a flash

156 FLAMMABLE HAZARDOUS MATERIAL

TABLE 8-3. Properties of Sulfur and Some Sulfur Compounds

	Flash Point	Ignition Temp.	Boiling Point	Melting Point	Flammable Limits (Percent by Volume in Air)		Vapor Density (Air=1.0)	Specific Gravity (Water=1.0)	Water Soluble	DOT Label Requirements
					Lower	Upper				
Amyl mercaptan $C_5H_{11}SH$	65°F 18°C	—	260°F 127°C	-104°F -76°C	—	—	3.6	0.8	No	Flammable liquid
Carbon disulfide CS_2	-22°F 30°C	212°F 100°C	115°F 46°C	-167°F -111°C	1.3	44.0	2.2	1.3	No	Flammable liquid
Dimethyl sulfate $(CH_3)_2SO_4$	182°F 83°C	842°F 450°C	370°F 188°C	-16°F -27°C	1.4	7.5	4.4	1.3	Very slightly	Corrosive material
Dimethyl sulfide $(CH_3)_2S$	0°F -18°C	402°F 206°C	99°F 37°C	-145°F -98°C	2.2	19.7	2.1	0.8	No	Flammable liquid
Hydrogen sulfide H_2S	— —	500°F 260°C	-77°F -61°C	-176°F -116°C	4.3	45.0	1.2	1.08	Yes	Flammable gas and poison
Sulfur	405°F 207°C	450°F 232°C	832°F 444°C	234°F 112°C	—	—	Varies	2.0	No	None
Sulfur chloride S_2Cl_2	245°F 118°C	453°F 234°C	280°F 138°C	-112°F -80°C	—	—	4.7	1.7	Decomposes	Corrosive material

point. Carbon has a boiling point so high that the solid combines directly with atmospheric oxygen without vaporization. White phosphorus also ignites in the solid state. Elemental sulfur finds yet another way: It melts at 234°F (112°C). Just like flammable liquids, molten sulfur can flow and spread fires, although only between 234°F and 392°F (112°C and 200°C). It has the peculiar property of increasing in viscosity as its temperature is raised. The color of liquid sulfur darkens, and stickiness increases up to 392°F (200°C). Above this temperature, the color lightens again, viscosity diminishes, and flowing starts in earnest. Molten sulfur has a flash point of 405°F (207°C) and, within a narrow range of 45 degrees F (25 degrees C), sulfur vapors can be ignited by a source. At 450°F(232°C), sulfur's ignition temperature, a source is no longer necessary. At 832°F (444°C), molten sulfur boils into a gas.

Fighting Sulfur Fires

With clouds of yellow-to-orange gases, molten dark brown pools, the choking odor of poisonous sulfur dioxide, and the blue flames of burning brimstone, a sulfur fire can be a fearsome sight. Small wonder this was considered a reasonable approximation of the fires of Hell, and sulfur was believed the stuff from which the eternal fires sprang. Fighting of sulfur fires should include the use of fog patterns that will cool the element below its ignition temperature and somewhat reduce concentrations of sulfur dioxide. Although there is some possibility of a steam explosion, sulfur is more than twice as heavy as water and not soluble. Water should not be trapped easily. While the careful use of water fog can be very efficient, avoid the use of straight streams. Not only will they increase the likelihood of a steam explosion, but they can also create clouds of sulfur dust that may explode. The ignition temperature of such a cloud is 75°F (42°C) lower than the solid, beginning around 375°F (191°C).

The toxicity of pure sulfur is quite low. The solid material has no action upon the skin, although sulfur *compounds* have been used in various ointments for skin ailments since ancient Egyptian times. Sulfur dioxide, sulfur's combustion product, is highly toxic, however, and molten sulfur can burn the skin badly. Wear protective clothing and use self-contained breathing apparatus.

The annual production of sulfur is well over 15 million tons, and it is an element of many uses. One is to vulcanize rubber. Before vulcanization, rubber is structurally weak and easily deformed, softened by heat, made stiff by cold, and soluble in many common liquids, including gasoline. Vulcanization modifies all of these undesirable properties. Sulfur forms a great variety of compounds: sulfuric acid, carbon disulfide, sulfur dioxide, hydrogen sulfide, and others. Pure sulfur can act as a fungicide in the lime and sulfur mixture familiar to gardeners. The main use of sulfur is in the production of the so-called "workhorse of industry," sulfuric acid, H_2SO_4. It has been said that one index of a nation's economy can be found in the amount of sulfuric acid it produces annually.

PHOTO 8–3. DOT corrosive label. This label, with black-and-white background and white lettering, is required on all shipments of corrosive materials. The corrosive placard is identical.

Along with saltpeter and charcoal, sulfur is one of the components of black powder. If sulfur comes into contact with a pure carbon material, such as charcoal, it can spontaneously heat and ignite. Otherwise, it has little tendency to do so. In storage, because of this property, sulfur should be kept separated from oxidizers and carbonaceous materials. Storage areas should be cool, well ventilated, and free from accumulations of dust.

An incident involving molten sulfur occurred in the town of Benicia, California. A truck carrying hot liquid sulfur at 325°F was crossing over a bridge when it hit a passenger car, lost control, crossed over the center divider, hit another truck, and overturned. Passengers in the cars involved were covered with the molten sulfur. Six firefighters, one police officer, and two toll-takers at the bridge toll plaza were overcome by the toxic fumes. Firefighters applied cooling water in an effort to solidify the sulfur so it could be scooped up with skip loaders. The driver of the truck was killed and two people were sent to the hospital with serious burns. An elderly male had third degree burns over fifty percent of his body and his wife had second degree burns over twenty-one percent of her body.

Sulfur Compounds

Amyl mercaptan, $C_5H_{11}SH$, is a liquid composed of a mixture of isomers. It is colorless to light yellow, has a very disagreeable odor, is a skin and respiratory irritant, and is used as an odorant for detection of natural gas leaks. It is shipped in 1- to 55-gallon drums (4 to 207 liters) with a DOT red label: flammable liquid. Foam, carbon dioxide, or dry chemical can all be effective on fires. (See Table 8–3 for other properties.)

Antimony pentasulfide, Sb_2S_5, often called red antimony, is an orange to yellow odorless powder that is combustible and easily ignited. It yields hydrogen on contact with strong acids, particularly concentrated hydrochloric, and it yields sulfur dioxide when burning. It is shipped in cans and fiber drums and is used as a pigment, in vulcanization and coloring of rubber. Use water and wear full protective equipment.

Dimethyl sulfate, $(CH_3)_2SO_4$, is a colorless liquid. If it touches the skin for even short periods, it can cause intense irritation several hours later. There is no odor or immediate irritation to give a warning, so it is very dangerous. Heavy exposures can cause death. Spilling of the liquid on the skin can also cause ulcers. Complete protective equipment should be worn. It is shipped with a DOT black and white label: corrosive material. (See Table and Photo 8–3.)

Dimethyl sulfide, $(CH_3)_2S$, is a colorless to straw colored flammable liquid with a disagreeable odor, shipped in drums and tank cars with a DOT red label: flammable liquid. It is very toxic, with high concentrations of vapor quickly reached because of its low boiling point. Use water spray, carbon dioxide, or foam on fires. Sulfur dioxide is the combustion product. Wear protective clothing and self-contained breathing apparatus. (See Table 8–3.)

Hydrogen sulfide, H_2S, is a colorless flammable gas with a strong offensive odor, shipped in steel pressure cylinders with a DOT red label: flammable gas, *and* a black and white label: poison gas. Typical cylinder pressure at 70°F (21°C) is 260 psi (18 atmospheres). Hydrogen sulfide is very toxic, and moderate concentrations can cause immediate death. Low concentrations are irritating to the eyes and respiratory tract. Major fire objectives: stop flow of gas, keep cylinders cool, dilute concentrations of this water-soluble gas with fog patterns. Wear self-contained breathing apparatus. (See Table 8–3.)

An incident involving hydrogen sulfide took place at the Texaco Refinery in Wilmington, California, on October 8, 1992. The explosion was recorded as a sonic boom by Cal-Tech seismometers in Pasadena, some twenty to thirty miles away. The initial blast damaged the on-site water system and water relay operations had to be initiated. Due to the toxic characteristics of hydrogen sulfide, 500 residents were evacuated in a two and one-half mile area as a precautionary measure. Since the fire was consuming the leaking liquid at approximately the same rate as it was leaking, the fire was allowed to burn itself out until shutdown of the system could be accomplished. This is just another example of the successful management of an emergency situation by knowing the inherent characteristics of a given flammable liquid.

Lead sulfocyanate, $Pb(SCN)_2$, is a white to light yellow crystalline powder, shipped in fiber and stainless steel drums that should be kept closed. It does not burn rapidly but decomposes in air to form carbon disulfide and sulfur dioxide. The dust is a possible explosion hazard. Use water and wear self-contained breathing apparatus. It is used in priming mixtures for small arms ammunition, safety matches, and the manufacture of dyes.

Sulfur chloride, S_2Cl_2, is an amber to yellow oily liquid with a penetrating odor. The fuming liquid is corrosive. The vapors are irritating to the eyes, lungs, and mucous membranes. Wear complete protective equipment. Sulfur chloride is shipped in iron drums inside wooden crates with a DOT black and white label: corrosive. It decomposes on contact with water. Uses include: sulfur solvent, poison gas manufacture, hardening of soft woods. (See Table 8–3.)

Sulfur dioxide, SO_2, is a colorless gas with a choking odor. It is highly irritating to the eyes and respiratory tract. Low concentrations can cause death. Its vapor density is 2.3, and its boiling point is 14°F (-10°C). It is shipped with a DOT green label: nonflammable gas. It is shipped in steel pressure cylinders containing 200 to 800 pounds (91 to 364 kg) of liquefied gas or in large tank cars. Typical cylinder pressure at 70°F (21°C) is 35 psi (2.3 atmospheres). It is used industrially as a bleaching agent, food preservative, fumigant, color restorative in aged grains, and in the manufacture of chemicals. It is the common combustion product of sulfur-containing materials. Major fire objectives: stop flow of gas, keep cylinders cool. Wear self-contained breathing apparatus.

CARBON DISULFIDE

If a mad scientist decided to create a super flammable liquid, what properties would he give the monster?

1. *A flash point and a boiling point so low that flammable vapors will generate in abundance at almost any temperature.* The combination of a flash point of -22°F (-30°C) and a boiling point of 115°F (46°C) will mean that vapors are produced early and often. If the material is heated even slightly, vapor pressure will be considerable.
2. *An ignition temperature so low almost any source of heat will be sufficient to explode its plentiful vapors.* The boiling point of water is about right. Although another compound discussed earlier, phosphorus sesquisulfide, has the same ignition temperature, it is a solid and tends to stay put. However, once free, vapors of a liquid will travel and seek out an ignition source. At 212°F (100°C), an ordinary light bulb, a steam pipe, a metal roof heated by sunlight, or any one of a dozen other sources of heat not ordinarily considered hazardous can be that ignition source.
3. *An extremely wide flammable range so that percentages too rich and too lean are almost impossible.* A range of 1.3 to 44 percent ensures

the explosive quality of almost every collection of vapor, whether caused by a small leak or a major spill. In addition, easily reached concentrations between 4 and 8 percent should explode with a maximum of violence.

4. *A high vapor density to make sure that vapors stay together until they find an ignition source.* A density more than twice as heavy as air, 2.2, ensures the same type of cloud that makes the LP gases so dangerous. Of course, the liquid and vapor should *not* be water soluble. This would make it too easy for firefighters to wash it out of the air with fog patterns. The liquid should be soluble in alcohol, ether, and benzene; none is recommended as a dilution agent in firefighting.

5. Finally, add a few minor touches: Make it so toxic that firefighters must have their vision partially obscured by masks. Make it so cheap and useful that industry will produce it in quantity. Make the liquid *mobile,* which means that it will run all over the place because it has little viscosity. Not a bad recipe for a monster.

Put all these ingredients together and they spell **carbon disulfide,** a liquid with a unique combination of properties. It is generally considered *the most dangerous of all common flammable liquids.* In a hazardous classification system where kerosene is rated at 40, a flammable liquid like ethyl alcohol at 70, gasoline at 95–100, and ethyl ether at 100, the Underwriters' Laboratories give carbon disulfide the gold medal rating of 110. Not content with that, they add a plus. Any 110+ liquid deserves a section of its own. (See Table 8–3.)

Small amounts of carbon disulfide occur naturally in coal tar or crude petroleum, but it is prepared on an industrial scale either by heating pure carbon (charcoal, coke, or coal) in a furnace, along with sulfur vapors, or by the interaction of sulfur and methane vapors. Because both end products, carbon disulfide and hydrogen sulfide, are useful, this process has become increasingly popular since the 1950s.

$$CH_4 \ + \ 2S_2 \longrightarrow CS_2 \ + \ 2H_2S$$

Methane   **Sulfur**   **Carbon**   **Hydrogen**
                                     disulfide   **sulfide**

Unfortunately, carbon disulfide has a range of properties that makes it valuable. It will dissolve phosphorus, sulfur, selenium, bromine, iodine, fats, resins, and rubbers. Toxic enough to be an insecticide or rodenticide, it is most effective as a fumigant at concentrations slightly above its lower flammable limit. The *major use* of carbon disulfide is in the manufacture of rayon and cellophane, although large quantities are employed to cold-vulcanize rubber, to manufacture carbon tetrachloride, and to make explosives.

Carbon disulfide is a colorless to pale yellow liquid with a strong disagreeable odor that has been poetically described as similar to that of decayed cauliflower, moldy broccoli, or rotten cabbage. The purest distillates are reported to have a sweet and pleasing aroma. High concentrations of vapor, whether sweet or foul, are narcotic and cause death by respiratory paralysis. One part in 250 in air (4,000 parts per million) is fatal in one to two hours. Lower concentrations can cause symptoms of intoxication, muscular weakness, sensory impairment, and unconsciousness. *It is poisonous by oral intake, inhalation, or prolonged contact with the skin.* Wear protective equipment and self-contained breathing apparatus.

Extinguishing a Carbon Disulfide Fire

How can fire from such a monster be extinguished? As stated previously, carbon disulfide is not soluble in water, but it has two other properties that help formulate a sensible plan of attack. First, it burns with a very low heat of combustion, less than half as hot as gasoline; second, the specific gravity of carbon disulfide is 1.3 times heavier than water. Not much, but these properties do offer some help.

How do ordinary firefighting agents work on carbon disulfide? Foam is ineffective because CS_2 vapors will penetrate the blanket and burn above it; the vapor pressure caused by a low boiling point is too much for foam. Carbon dioxide is not too efficient for a different reason. Complete extinguishment requires an atmosphere of 55 percent CO_2, twice as much as with gasoline. Here the 44 percent upper flammable limit of CS_2 is in operation. Dry chemical is not a world beater, either. Carbon disulfide vapors re-ignite easily because of their low ignition temperature. Dry chemical has little cooling effect on hot surfaces. In order to have any chance of success on a carbon disulfide spill fire, dry chemical must be employed in conjunction with cooling water spray.

This leaves us with our old reliable, water. Because of its specific gravity and low heat of combustion, a flaming tank of carbon disulfide is most likely to be extinguished through the use of water. A fairly close approach, from upwind of an involved container, favored by a very low heat of combustion, may be possible. If so, **blanketing,** floating a layer of water over the top of a high-specific-gravity liquid that is not soluble in water, may do the trick. But blanketing must be done gently. An open hose-butt is the best applicator because it will deliver quantities of water without agitating the surface unduly. A thick enough layer of water over a carbon disulfide fire will extinguish it.

Special Storage Requirements

A material such as carbon disulfide must have special storage requirements. Here, indeed, is where an ounce of fire prevention is worth a ton of fire-extinguishing agents.

Store carbon disulfide containers in an isolated, preferably noncombustible building in which they may be safeguarded against injury. The area should have no heat, direct sunlight, hot pipes, or electrical lighting fixtures.

Any artificial light should shine through glass ports. Protect against static electricity. Spray drums with water during hot weather to reduce vapor pressures. Provide floor-level ventilation for indoor storage. (Construct large tanks to allow for underwater storage or over concrete basins large enough to hold all the contents of the tank in addition to the water.) Keep large containers blanketed with water or inert gas at all times. Use wooden measuring sticks to check tanks, not spark-producing metals. Never dispose of carbon disulfide by dumping on the ground or into sewers, but always by burning in a safe location. It should be transferred by pump, or displaced by an inert gas or by water, rather than by air pressure.

Finally, CS_2 is shipped in small gas or metal containers (1- to 5-pound bottles or cans) inside *ordinary wooden or cardboard boxes,* in 5- to 55-gallon drums, or in tank cars up to 7,000 gallons. The DOT requires a red label: flammable liquid. Transportation is of more than passing interest, for carbon disulfide was the liquid that caused a fire in the Holland Tunnel. The following description is condensed from the May 13, 1949, report of the AIA.

At about 8:45 A.M. on Friday, May 13, 1949, a large trailer truck, loaded with 80 fifty-five-gallon drums of carbon disulfide entered the Jersey City side of the Holland Vehicular Tunnel that runs under the Hudson River to New York City. Commercial traffic was particularly heavy that morning and moving slowly. Five trucks, including the carbon disulfide unit, entered the tunnel at about the same time. About 350 feet behind them was another group of five trucks. [See Figure 8–3.]

Three minutes later, a patrolling officer saw a truck, apparently stalled in the tunnel, some 2900 feet from the New Jersey entrance. As he ran toward the truck, a loud blast occurred and he was met by two men. He guided them to safety. Another tunnel officer, noticing dense fumes, smoke, and no emerging traffic, sent the alarm.

Automatically, signal lights behind the emergency turned red, halting incoming traffic; signals ahead turned amber, shifting traffic to the right-hand lane and slowing it down and opening the left lane for emergency equipment. The Port Authority crew, entering the tunnel from the New York end, which was free of traffic, discovered a large ball of fire in the tube. Protected by all-service masks, they were able to connect a 1 1/2-inch hose to a standpipe outlet about 75 feet from the first burning truck, knock down the fire in trucks 1 and 2, and advance on truck 3 when they noticed flaming liquid running down the gutters on both sides of the tunnel. The first rescue company from the N.Y. Fire Department arrived at this point and took over operation.

They found a very serious situation: a number of trucks on fire, extreme heat, bursting carbon disulfide drums, toxic fumes, and minimum visibility because of heavy smoke. In addition, broken circuits had put the tunnel lights out and damage to the tunnel ceiling threatened a possible collapse of the tube itself. Fireboats were positioned above the tunnel to watch for tell-tale bubbles, showing a leak.

So intense was the fire that it spread through 350 feet of empty tunnel to the second group of trucks, near the Jersey end. All the wall surfaces and ceiling slabs of this section were demolished; large sections of concrete were left hanging by their reinforcing bars. The roadbed was completely covered with broken concrete and tile and the trucks involved were reduced to twisted heaps of metal.

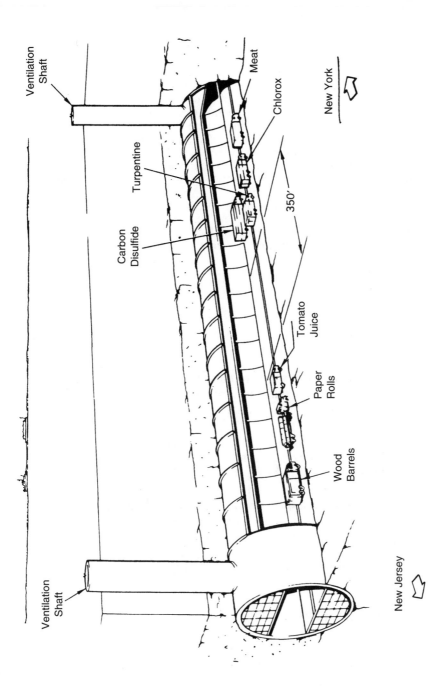

FIGURE 8-3. Holland Tunnel fire.
Source: American Insurance Association.

Ventilation Shaft

Ventilation Shaft

Meat

Chlorox

Turpentine

Carbon Disulfide

Tomato Juice

Paper Rolls

Wood Barrels

New York

New Jersey

350'

Until 125 stalled cars were removed, the Jersey City Fire Department, moving from the other end of the tunnel, had to hand-carry hose to the standpipes closest to the fires in the second group of trucks. In spite of heavy fumes, they advanced quite rapidly, putting out the nearest truck fires one at a time.

By nightfall, in spite of a re-flash of carbon disulfide fumes that was quenched by foam, the fire was under control. Cleanup required the removal of over 650 tons of debris from the tunnel. There were no fatalities but a great many firemen were hurt. The major cause of injury was smoke and gas inhalation; most of the first men from the NYFD, several of the Port Authority crew, and ten of the Jersey City men had to be treated. Wartime civil defense gas masks, pressed into use, were ineffective against the toxic gases in the tunnel.

All this damage was caused by a single truckload of carbon disulfide, showing the untold potential hazards in our normal day-to-day highway transportation of dangerous chemicals. By sheer luck, passenger cars and buses, including three buses full of children, were not in the tunnel when the fire began.

SUMMARY

Solids find many ways to burn.

1. An ordinary flammable solid produces vapors at its ignition temperature that are consumed as they are produced.
2. A solid like camphor or naphthalene passes directly from a flammable solid to a gas. Because the ignition temperature is still higher, these vapors have time to build up to their flash point.
3. Solids like white phosphorus and carbon can ignite in their solid state. When, and how, they do so varies.
4. Like sulfur, a flammable solid can melt. The liquid then produces vapors and reaches a flash point. Flammable solids can have flash points to go along with their ignition temperatures. Ignition can occur before or after the solid has melted into a liquid or boiled into a gas. The flash point can come either in the solid or liquid state. (See Table 8–4 for examples of this versatility.)

REVIEW QUESTIONS

1. What does sublime mean?
2. Name two examples of solid materials that have a flash point.
3. What is an allotrope?
4. Why is the use of water not advised on a fire involving a pure carbon material?
5. What are the ignition temperatures of white and red phosphorus?
6. How is white phosphorus shipped and stored?
7. Why should straight streams be avoided when fighting sulfur fires?

TABLE 8-4. Properties of Selected Flammable Solids

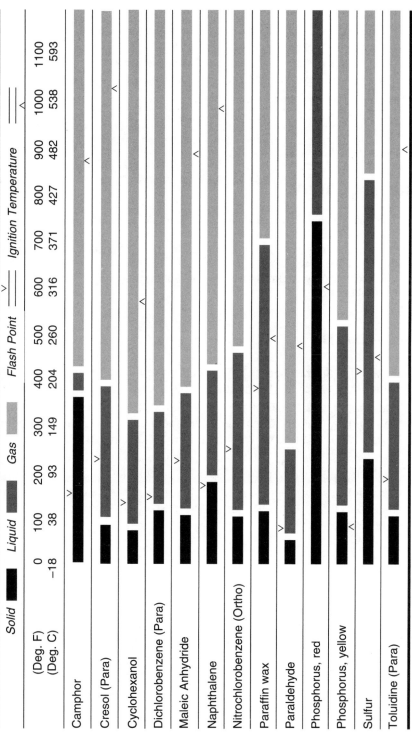

8. Discuss the toxicity hazards of dimethyl sulfate.
9. What effect will fog patterns have on concentrations of hydrogen sulfide gas?
10. Discuss the firefighting of carbon disulfide.
11. When does a Class A solid produce flammable vapors?
12. Name two factors that influence the flammability of ordinary solids.
13. Name three types of carbon black.
14. What is the recommended extinguishing agent for a phosphorus fire?
15. Of the two phosphorus compounds discussed in this chapter, which has the lowest ignition temperature? Which is most reactive with water?

BIBLIOGRAPHY

"The Holland Tunnel Chemical Fire,"*American insurance Association Report,* May 13, 1949.

"Carbon Disulfide," *Handbook of Industrial Loss Prevention,* Factory Mutual Division, Chapter 52, 1967.

"Chemical Safety Data Sheet SD-74—Sulfur," Chemical Manufacturer's Association.

VISUAL AIDS

8:Hydrogen Sulfide. Detrick Lawrence Corporation, The Emergency Film Group; 225 Water St.; Plymouth, MA 02360. (28 minutes, VHS format).

Phosphorus and the First Responder. Phosphorus Mutual Assistance Companies (PERT); Chemical Manufacturers Association, Lending Library, 2501 M St., NW, Washington, DC. (22 minutes, VHS format).

Hydrogen Sulfide Fire—Wilmington, California. American Heat Video Production, 240 Sovereign Ct. Ste. C, St. Louis, MO, Vol. 7, Program 5.

Benicia-Martinez Bridge—Molten Sulfur-Truck Accident, Hazardous Materials Library, FEMA Region IX, Presidio of San Francisco, FL5-6-19A, VHS format.

Combustible Metals

The properties of eight gaseous elements have been previously discussed: highly flammable hydrogen; oxygen, the combustion supporter; and some more or less inert gases—nitrogen, helium, argon, neon, krypton, and xenon. A large part of the previous chapter emphasized the differences between three elemental flammable solids: carbon, sulfur, and phosphorus. All these elements have one thing in common: None is a metal. This is more unusual than it may seem, because most of the other elements are metals. Of the 106 or so elements, only 22 are nonmetals. Half of this total are mentioned above; five more nonmetals will be met in a section on halogens: fluorine, chlorine, bromine, iodine, and astatine. (Astatine and radon, another inert gas, are both radioactive.) These 17 elements make up the major part of the nonmetallic group. Several others, including boron, selenium, arsenic, silicon, and tellurium, have properties that place them somewhere in between metals and nonmetals. As a result, they are often called *metalloids*, or semimetals.

This leaves us with the properties and hazards of 81 metals to learn. Although many are so rare as to rate only a brief mention, there is a lot of ground to cover.

First, a series of true statements can be made about metallic elements in general, although there are some exceptions to them.

1. Most metals are lustrous. They are shiny, especially when polished.
2. Most metals are good conductors of heat and electricity at ordinary temperatures. Most nonmetals are insulators. There is also a variety of materials that have the qualities of both conductors and insulators. Examples are carbon and silicon.
3. Most metals are malleable. They can be hammered or formed into sheets or otherwise shaped.

4. Most metals are flexible. They will bend under stress and return to their original shape when the tension is relieved.
5. Most metals are ductile. They can be drawn into wires or threads.
6. Most metals have tensile strength. They resist being pulled apart lengthwise.
7. Most metals are silver-white or grayish. Two familiar exceptions are copper and gold.
8. Most metal powders are dark gray or black. Notable exceptions are aluminum, beryllium, and magnesium, which are silvery.
9. Most metals are solid at ordinary temperatures. Of course, there is mercury, and a rare metal, gallium, which melts at 85°F (29°C).
10. Most metals will combine with oxygen to form oxides. The most costly oxide is formed by iron. Millions of dollars are spent each year for rust preventatives.

The names given to elements are supposed to reveal what they are. Recently discovered metals have names ending in *-um* or *-ium*, non-metals in *-n* or *-ne*. Nonmetals conform to the rules pretty well, although sulfur, phosphorus, and arsenic go their own ways. Tellurium and selenium have metallic names, although their properties do not always conform to what their names promise. A glaring exception is helium, an inert gas, which was discovered on the sun and was thought for a time to be the "sun metal." Such metallic nonconformists as manganese, cobalt, nickel, silver, lead, copper, zinc, iron, mercury, and antimony were named long before the rules were developed.

HAZARDS OF METALS

Metals can be toxic or corrosive. No fewer than 20 metals are radioactive. And most metals have another hazard that is important: They will burn. The flammable potential of 81 different metallic elements varies about as widely as would the same number of hydrocarbons. For example, one of the reasons why precious metals are so prized is that they resist oxidation and thereby retain their luster. (If you doubt the value of this property, give the love of your life a ring of brass and note the comments when the brass oxidizes and his or her finger turns green. A precious metal that steadfastly refuses to oxidize will also be stubborn about igniting readily. Other metals are less resistant, and any metal that reacts with oxygen and releases heat is a candidate for ignition, if conditions are favorable.

Form and Shape

Perhaps the single most important condition that regulates the combustibility of a metal is its form and shape. In this respect, a metal is no different from any other combustible solid, be it charcoal, wood, or plastic.

PHOTO 9–1. DOT oxidizer label. This yellow label with black lettering is required on containers in which oxidizing materials are shipped. A similar oxidizer placard is required when a shipment via motor vehicle or rail car carries a gross weight of 1,000 pounds (454 kg) or more.

This fact will be repeated a dozen different ways during this chapter as the hazards of individual metals are discussed. First, a few more general comments may clarify the picture.

1. Some metals, difficult to ignite in a solid massive form, burn readily as thin sheets or shavings. Steel wool will burn, as will the aluminum foil ignited by an electric spark in a photographic flash bulb. As the particles become finer and finer, the ignition temperature of the metal lowers. Surface areas grow, and so does the hazard.

2. Dust clouds of such seemingly noncombustible metals as copper and tungsten are potentially explosive. What is true for these metals is even more likely for powdered titanium, a metal that burns readily. An industrial establishment that stores or creates metal powders or dusts by some process has the ingredients for an explosion on its premises.

3. Some metal powders can ignite spontaneously in air. When a material does this at ordinary temperatures, it is said to be **pyrophoric**. Pyrophoricity is a property shared by many flammable metal powders and dusts: magnesium, calcium, sodium, potassium, zirconium, hafnium, a special form of nickel, and several others. A layer of pyrophoric powder can ignite after a few minutes' exposure to atmospheric oxygen. It can ignite immediately after dispersion in air

as a cloud. Metal powders have been known to burn in pure carbon dioxide, nitrogen, and argon, or even under water.

4. Some of these metal powders, when moist, are capable of producing an explosion more violent than one caused by nitroglycerin or TNT, and they can explode in the presence or the absence of air. What is most significant is that this heating, igniting, or exploding has been known to occur with no warning or source of heat.

5. Many metals will react violently with water in one way or another no matter what their form, massive or finely divided. Some elements, the alkali metals (lithium, sodium, potassium, rubidium, and cesium), will explode when water comes into contact with them. The metals discussed here react most vigorously with water when they are heated or burning.

Combustion Environments

As mentioned, metals can burn in some surprising environments. But combination with oxygen is still the most common form of combustion. Many metals are not too particular about where they obtain their oxygen. They will decompose water, burning with the aid of the oxygen in the H_2O molecule and releasing flammable hydrogen gas. Metals can also react with other chemicals to release hydrogen. Chromium, cobalt, iron, manganese, nickel, and zinc will react with acids; aluminum, titanium, and zirconium react with caustic alkalis.

The use of water streams on such metals merely intensifies the combustion process rather than cool the situation. Metals have been known to burn more violently in contact with steam than with pure oxygen. Small pieces of many metals burn more intensely when wet than they do when dry.

The use of water on some metals has another hazardous possibility. Although some metals melt and burn at the same time (magnesium, for example), the dangers created by a molten metal are not necessarily confined to those metals that are flammable. When any metal contributes molten material to a fire situation, there is a good chance of a severe or even disastrous steam explosion if water is used unwisely. (See melting temperatures in Table 9–1.)

Firefighting methods must be considered carefully when metals, particularly metal powders, are involved. All of the common extinguishers—water, foam, dry chemical, carbon tetrachloride—and some of the inert gases may either stimulate the burning process or cause an explosion. A great deal of thought about alternatives and consequences must go into our sizeup and decision, especially if the fire also involves other types of combustibles, such as flammable liquids. The extinguishing agent recommended for a flammable liquid may cause a disaster if it meets a metal powder.

TABLE 9–1. Temperatures at Which Metals Become Molten

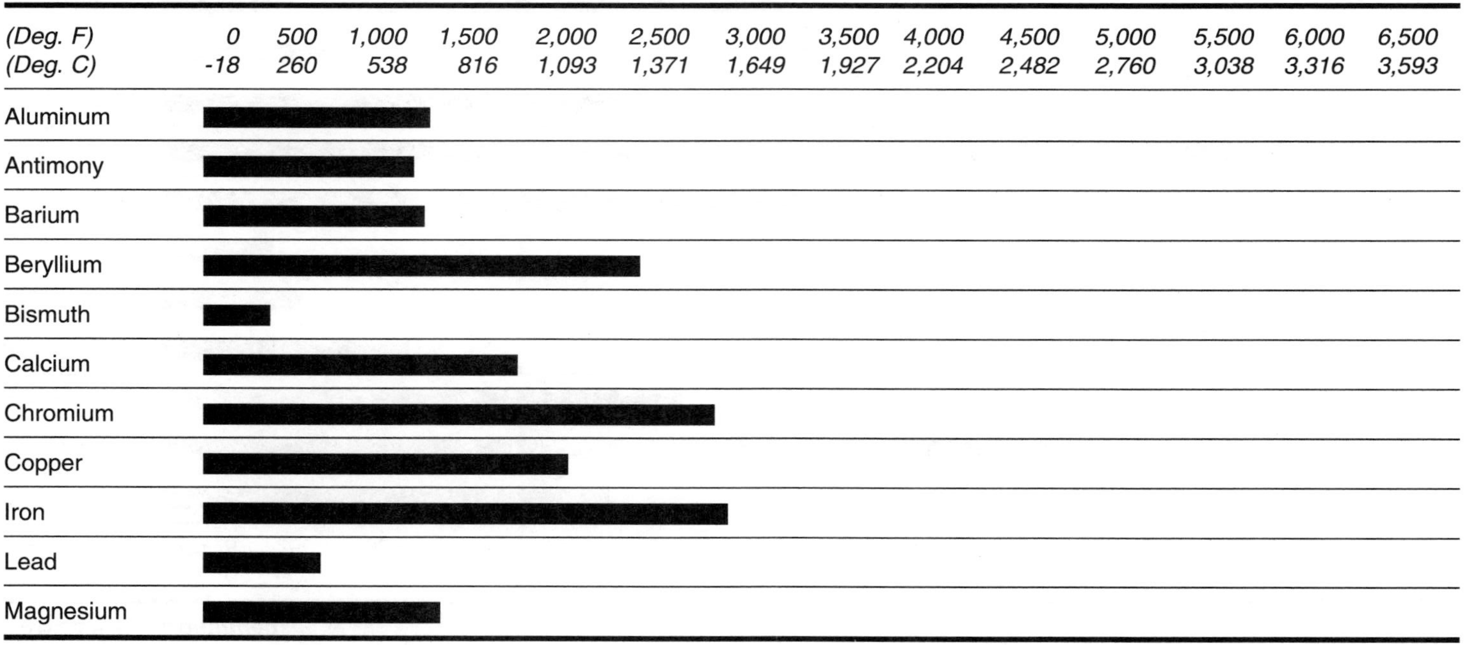

(Deg. F)	(Deg. C)
0	-18
500	260
1,000	538
1,500	816
2,000	1,093
2,500	1,371
3,000	1,649
3,500	1,927
4,000	2,204
4,500	2,482
5,000	2,760
5,500	3,038
6,000	3,316
6,500	3,593

Aluminum

Antimony

Barium

Beryllium

Bismuth

Calcium

Chromium

Copper

Iron

Lead

Magnesium

TABLE 9-1. *Continued*

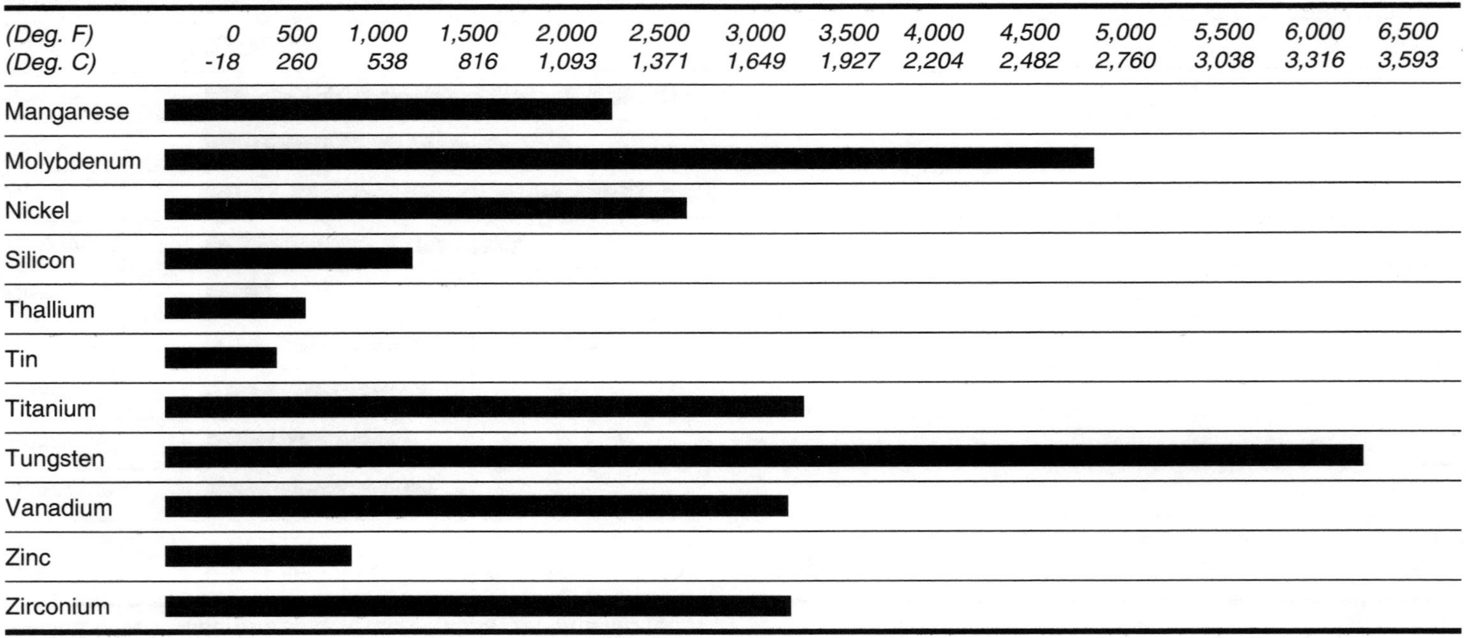

(Deg. F)	0	500	1,000	1,500	2,000	2,500	3,000	3,500	4,000	4,500	5,000	5,500	6,000	6,500
(Deg. C)	-18	260	538	816	1,093	1,371	1,649	1,927	2,204	2,482	2,760	3,038	3,316	3,593

Manganese
Molybdenum
Nickel
Silicon
Thallium
Tin
Titanium
Tungsten
Vanadium
Zinc
Zirconium

A closer look at specialized metal extinguishers will take place later in this chapter, but even they are not entirely foolproof. Inconsistencies may pop up. Recommended methods of extinguishment may work beautifully on one metal fire only to fail in another, although the situation seems identical. The same metal from different sources can react differently to a particular extinguishing agent. A massive piece of pyrophoric metal can ignite at widely varying temperatures. One time it will ignite while resting on a piece of dry ice; another time it will fail to catch fire when heated to a melting point far above its normal ignition temperature. Much has still to be learned about the pyrophoricity and flammability of metals, even those as common as magnesium.

MAGNESIUM

As far as industry is concerned, magnesium is one of the metals of the future. It is moderately hard and fairly strong. When alloyed with aluminum, zinc, and manganese, it is easy to machine on a lathe and work. Most important, magnesium is very light, less than twice as heavy as water (1.74), only 65 percent as heavy as aluminum, 25 percent as heavy as iron. There are many uses for such a lightweight metal: in aircraft, automobiles, railway cars, buses, truck bodies, furniture, ladders, and luggage, to name a few. Magnesium alloys are even beginning to find some structural employment.

There is some reason for describing magnesium as the "silvery metal from the sea." Each cubic mile of ocean contains about 6.5 million tons of magnesium. While this is a lot of metal, the amount of water is far greater. Nevertheless, there is an economical extraction process. Seawater and slaked lime (often obtained from oyster shells) will yield magnesium hydroxide. This milk of magnesia reacts with hydrochloric acid to form magnesium chlorate, which, in turn, is split by electrolysis into pure magnesium and chlorine gas. The chlorine is recovered to make more hydrochloric acid. This method and more conventional mining produce an annual total of well over 200,000 tons.

Hazards of Magnesium

To the firefighter, the future of the magnesium industry is less important than the problems the metal brings now. Although inhalation of magnesium fumes or magnesium dust may cause some irritation of the throat and lungs, and the brilliant white light of burning magnesium has been known to hurt the eyes, it is not its toxicity but the flammable potential of magnesium that produces the concerns. Uses of the metal and its powders reflect this property: pyrotechnics, photographic flash powders, military flares, and incendiary bombs. Magnesium, combined with white phosphorus in a bomb, makes a fearsome device: The same water that controls the phosphorus causes flareups of the burning magnesium.

Magnesium melts at 1,204°F (651°C) and boils at 2,048°F (1,120°C). The ignition temperature of a solid chunk is close to this melting point. Fine shavings and loose scraps will ignite at less than 1,000°F (538°C); powder, another hundred degrees F lower. Some magnesium alloys catch fire at less then 800°F (427°C). What is significant about these temperatures is that all of the common ignition sources reach these temperatures. This is especially true with respect to magnesium powder, whether deliberately produced for some purpose or unavoidably created by the grinding operations of a machine shop. As can be expected, heavier pieces of magnesium are far more difficult to ignite and are often stored in the open like other metals. As a wooden log will transmit heat away from a point source of ignition, so will a metal with a higher rate of heat conductivity. Although an entire piece of solid magnesium must be raised to its ignition temperature before it ignites, this is easily done in a fire. In addition, magnesium is one of those metals that melt as they burn. Larger pieces will contribute far more molten metal to the situation than an equivalent weight in smaller pieces because the smaller pieces are more likely to be completely consumed.

When heated, the magnesium molecule not only readily combines with oxygen, it does so with very dangerous consequences. It will burn in pure carbon dioxide, combining with the O_2 in CO_2, making the use of a carbon dioxide extinguisher ineffective. Dry sand is also ineffective. Oxygen is extracted from the silicon dioxide. If the sand is wet, the burning metal may produce steam and blow the sand pile apart. Burning magnesium will use asbestos as an oxidizing agent and continue burning even if the asbestos cuts off the air. The hunger for oxygen in burning magnesium is so great that it will burn more violently on a concrete floor than on a wooden one because the metal reacts with the oxygen in the concrete's aggregate. In spite of this, magnesium should not be stored on wooden floors.

Furthermore, if oxygen is not available, magnesium will "make do" with whatever is on hand. It will decompose carbon tetrachloride violently, getting at the chlorine atoms to support combustion. Even "inert" nitrogen gas supports combustion. Magnesium combines with it, forming magnesium nitride, and will burn in a pure nitrogen atmosphere. Argon, helium, and neon still remain aloof and will extinguish the burning metal.

Water on Magnesium Fires

In spite of what has already been mentioned, there is some disagreement over the use of water on burning magnesium. The Magnesium Association in its bulletins to metal-finishing shops states flatly: "Never use water, fog spray, foam, common gas, or liquid fire extinguishers on magnesium."

But other authorities, such as the NFPA (National Fire Protection Association) partially disagree with this absolute ban on water.

Magnesium, if hot enough, reacts violently with water with liberation of hydrogen, but the idea has been overdone that water should never be used on any kind of a magnesium fire.

When experts disagree, we must look for ourselves. There are two basic methods of applying water to a fire of any type: straight streams or fog patterns. The method used on magnesium, if either, depends upon sizeup. What amounts of metal are involved? What form is it in—powder, pieces, or large chunks? What is the extent of the fire? How long has it been burning? What is the type of building construction? How and where is the metal stored? What are the exposures?

If magnesium powder is present, it can be blown into the air by a straight stream. A dust explosion of tremendous intensity is possible. Straight streams must be used with extreme caution if magnesium dust is present. When there is a fire in a large quantity of small pieces, a straight stream can penetrate into the pile and cause an explosive scattering of molten and burning particles. The danger to personnel is evident. Although this scattering is often mistaken for a hydrogen explosion—the residue of the H_2O molecule after the oxygen is pirated away for combustion—most often it is simply a more intense version of an old-fashioned steam explosion, caused by the trapping of water.

If molten metal is involved, however, fog is much safer, for it will flash into steam immediately. But fog will not do the job properly. A small magnesium fire in chips and pieces simmers rather quietly if there is no moisture present. If water is applied, particularly water fog, the burning metal flares up amazingly into a hot, blinding white flame. Those who have seen experiments in which oxygen is deliberately applied to a burning combustible have seen a reasonable approximation of this intensive acceleration. Literally, the same thing is happening. Water is being decomposed by the reactivity of magnesium and the chemical attraction to combine with oxygen. Only large amounts of water can overcome this increase in burning rate and lower the temperature of the metal below its ignition temperature. This is particularly true when massive pieces are involved, the pieces most likely to produce molten metal.

To summarize, fog patterns are much more likely to accelerate the burning rate of magnesium, but there is less chance of a steam explosion. A straight stream can stir up an explosive dust cloud. There is more likelihood of a steam explosion if molten metal flows over pools of water we create or if water is trapped beneath the surface of a pile of chips. Yet a stream produces more water in less time, and we must have this volume of water to cool the metal below its ignition temperature and the temperature at which it combines so avidly with oxygen.

Modern methods call for a compromise between fog and straight streams, extinguishing and cooling the burning metal with a stream at lower pressures that will break up into drops over the fire. These coarse drops will not violently accelerate a magnesium fire the way fog will, but they will flow over the metal and cool it. After this, coordinated hose streams can be worked into the fire. Small but well-advanced magnesium fires have been extinguished by this method in a short time. If large quantities of metal are involved, it may be necessary to use large streams from a distance.

> The problems of a big magnesium fire were never more evident than in the City of Industry on December 4, 1990. The Los Angeles County Fire Department was called to the scene of a fire involving magnesium chips stored in a warehouse. The chips were raw materials used in the manufacture of airplane engines and parts. The fire started when magnesium metal chips fell on an exposed electrical fire, causing ignition. Plant employees attempted to extinguish the fire before calling the fire department. The initial attack centered around protecting exposures and not applying water to the fire directly. The igniting chips would explode much like ammunition shells. Once enough resources were amassed at the scene, it was decided to apply heavy master streams directly to the fire to accelerate it. At this point, 5,000 gallons per minute of water was being applied. This decision was made only after careful consideration of what the consequences would be. The fire was under control approximately six hours after the arrival of the first fire department units. As stated previously, knowing the characteristics of a material, applying sound fire ground strategy and tactics, and making conscious decisions only after analyzing the possible consequences of those actions is the only way to successfully handle incidents involving flammable hazardous materials.

Sprinklers

But why not prevent the formation of molten magnesium in the first place? Large pieces of the metal must burn for a time before molten pools will form. If extinguishment or control is immediate, these pools may not be created. An automatic sprinkler system above magnesium storage would be invaluable. The NFPA allows the use of automatic sprinklers in storage areas containing heavy and light castings of magnesium where combustible cartons, crates, or packing materials are present.

Time can be of the essence in magnesium fires. Sprinkler systems can perform a vital function by holding down the extent of the fire until the arrival of the fire department. If a magnesium fire can be caught while it is still small, several alternative courses of action may be available.

Dry Powders

Two types of dry powders have been approved for use on dry or oily magnesium chips, turnings, and castings. (The phrase "dry powder" refers to metal extinguishing agents. If a dry extinguisher is designed for such purposes as flammable liquids, it is officially called a "dry chemical.") One of them, **Pyrene G-1**, is a graphite powder that conducts heat away from the burning metal, with a possible additive that turns into a gas when heated and helps exclude oxygen. The other, **Ansul Met-L-X**, is a salt (sodium chloride), with additives to prevent caking, combined with a plastic that melts and fuses the powder into a solid cake over the fire. Met-L-X will cling to vertical surfaces, a helpful property when larger pieces are involved. Both of these powders are noncombustible and nontoxic. They will not increase the combustion rate of magnesium if we run out prematurely. However, firefighters must get close to the fire before they can use them. The graphite must be applied by a hand scoop or shovel. The range of the sodium chloride extinguisher is less than 10 feet (3 meters), and care must be exercised not to blow burning metal around with the force of the ejected powder. If possible, good training should include practice with both of these powders.

Cooling water cannot be used on the magnesium fire in conjunction with these extinguishers. Not only would the flareup prevent close approach, but the water would destroy the cake trying to form over the burning metal. This cake should be an inch or so in thickness. Once it is in place, leave it alone. It will take some time for the metal to cool down.

Other possibilities exist for fighting a small magnesium fire. If the metal is on a combustible floor, put down a two-inch layer of powder, shovel the metal on top, and cover it with more powder. More sensibly, if a clean, dry drum is available, make a magnesium sandwich inside it and carefully haul the drum to a safe place. A few burning chips can be quickly handled by dropping them in a bucket of water. (These dry powder extinguishers have not been approved for magnesium powder by the Underwriters' Laboratories.) If the metal powder is exposed to a fire, try to separate the two before the powder becomes heated. Be careful about the use of water. Remember that *damp* magnesium powder can heat, generate hydrogen, and possibly explode.

Storage and Use of Magnesium

The Magnesium Association makes several recommendations for safe machining and grinding. Keep its advice in mind when making inspections.

When machining magnesium, take heavy cuts with sharp tools, never permit a tool to rub on the work, and always back off the tool when the cut is finished. When taking fine cuts, use a mineral oil sealant, *not* a water-soluble oil. Sweep up chips frequently and store in covered metal containers.

When grinding magnesium, it is "magnesium only" for a grinder unless the wheels are thoroughly cleaned before changing metals. All dust must be carefully cleaned up and not allowed to accumulate on clothing, surrounding floor areas, beams, or sills. Use a wet dust collector (the dust is caught by a water spray as soon as it is formed and settles as sludge) for extensive grinding. Use the collector for magnesium only. Remove sludge regularly, preferably at least once a day. Place in covered, vented con-

PHOTO 9–2. DOT dangerous-when-wet label. The label is blue with black lettering and symbols.

tainers and remove to a safe location for immediate disposal. Partially dry sludge is highly flammable. Dispose of the wet sludge by burning outdoors in an isolated area on a well-drained layer of fire brick, free from previous residue. Place dry combustible refuse, paper and wood, on top of a 3″ to 4″ layer (7 to 10 cm.) of sludge and ignite the paper in a safe manner to avoid burns from a possible hydrogen flash. Observe burning from a safe distance upwind.

Different forms of magnesium should be separated in storage. Finely divided pieces should be stored in noncombustible buildings with the finest in containers protected from sources of ignition, moisture, and contamination by halogens and acids. Large pieces may be stored out of doors.

Magnesium is shipped in all forms, from large ingots down to the powder in closed metal or cardboard containers. *The scrap and powder must carry a DOT red label: flammable solid, and the dangerous-when-wet label.*

BERYLLIUM

Magnesium is the most common of a group of six elements called the alkaline earth metals. Other members of the family are beryllium, calcium, barium, strontium, and radium. Radium, of course, is radioactive.

Beryllium has many attractive properties that can be extremely useful. It resembles aluminum, not only as a silvery metal or powder but also in many of its properties. Beryllium is the lightest hard metal known, much harder than magnesium, and its specific gravity is only slightly higher: 1.85. Copper and aluminum, when alloyed with beryllium, gain great elastic strength. Such an alloy can be bent back and forth thousands of times without breaking, the perfect metal for a lightweight spring.

Coupled with its strength and low weight, the high melting point of beryllium, 2,400°F (1,316°C), makes it useful in rocket nose cones. It takes a hot fire to melt a large solid piece. Although beryllium will oxidize in air, it is the least reactive of the alkaline earth metals. It is much less flammable than magnesium and is even used to decrease the flammability of molten magnesium and aluminum. Active at high temperatures, beryllium resists oxidation at normal temperatures because it forms a skin of stable oxide. As with most metals, the powder can form explosive mixtures with air.

Beryllium Hazards

Beryllium is extremely toxic. If it enters the body through inhalation of dust or fumes, or through a break in the skin, fatal poisoning can result from a single short exposure to incredibly low concentrations of the element or its compounds. Symptoms of poisoning include: skin irritation, eye burns, ulcers that will not heal, coughing, shortness of breath, and loss of appetite and weight. These symptoms can be delayed from three months up to six years. Death rate after an exposure is about 25 percent. For a time, fluorescent tubes contained a compound called beryllium zinc silicate. After World War II, this compound was replaced by another, far less toxic, but an ancient tube can still be around. Consider all obviously old tubes suspect and treat them like poison; they may well be.

Beryllium Powder

The powder is shipped under a DOT poison "B" label in steel and fiber drums. It should be kept away from air, acids, and moisture by storage in tight containers, preferably under an argon atmosphere. Firefighting procedures include the use of a dry powder. Water is not recommended. More important, personnel *must* wear complete protective equipment and self-contained breathing apparatus. After a fire involving beryllium or one of its compounds, all personnel must bathe completely, with particular attention to fingernails and hair. Clothing should be washed separately from uncontaminated clothing. Equipment must be washed down thoroughly.

CALCIUM, BARIUM, AND STRONTIUM

Calcium is a silvery metal that rapidly tarnishes to bluish-gray in air. It is softer, lighter, and more reactive than magnesium. It will react rapidly with warm water to produce a stream of hydrogen bubbles, but this reaction generally does not produce enough heat to ignite the gas. (Calcium will detonate, however, on contact with alkaline hydroxides.) When finely divided, calcium will ignite in moist air and burn with a red flame. It burns rather quietly in air without creating molten metal. Its melting point is 1,562°F (850°C). The burning metal produces highly irritating fumes of calcium oxide, better known as quicklime. The solid metal can react with the moisture of the skin and eyes to form calcium hydroxide, also caustic. Wear protective equipment and self-contained breathing apparatus; brush calcium dusts off equipment promptly. None of the common extinguishers is recommended. Water, foam, and halogenated hydrocarbons may cause an explosion or strong reaction. Dry chemical and carbon dioxide are inefficient. Met-L-X, Pyrene G-1, or soda ash is effective.

Calcium is shipped under a DOT red label as a flammable solid, a dangerous-when-wet label in airtight cans, well-stoppered bottles, and sealed drums. It is often stored under kerosene or naphtha. Containers should be protected from damage or high temperatures and kept away from areas where water may be used normally or in an emergency. An isolated, noncombustible building is recommended for quantity storage. Although many calcium compounds are common and useful, the pure metal has a limited employment in metallurgy, to harden lead, and in vacuum tubes.

Most of the fire properties, reactions, and storage requirements of barium and strontium are similar to calcium. Barium is silvery white. It melts at 1,300°F (704°C). The pure element is not commercially important. All water-soluble barium compounds are poisonous.

Strontium is a soft, silver-white metal that turns pale yellow in air. It melts at 1,386°F (752°C). There are no commercial uses for the pure metal. Both barium and strontium burn in air with a brilliant flame. Strontium chlorate, nitrate, and peroxide are all oxidizers and require DOT yellow labels with black lettering.

TITANIUM

The story of titanium is that of an ugly metal duckling turning into a valuable swan. For a long time after its discovery, titanium was just another laboratory curiosity, one of those silvery metals in the back pages of chemistry textbooks. Although it is one of the ten most abundant elements on Earth, it is difficult to produce in quantity because it is spread about too evenly. It would have been more convenient to have it piled up someplace where it could be mined. Even when the metal was isolated, it had a disappointing property—extreme brittleness.

Then, a method using molten magnesium was developed that could prepare titanium economically in a really pure form. It turned out to be tough, malleable, shock-resistant, and so hard it scratched steel. The weight of titanium is halfway between those of iron and aluminum. But it is four times as strong as aluminum and stands up under heat much better. In fact, pound for pound, titanium is stronger than any metal, including steel. (Before you rush out and sell your steel stocks, remember that a rod of steel one inch in diameter is much stronger than the same size rod of titanium. Of course, it also weighs a lot more.) Titanium also proved to be resistant to many types of corrosion, including salt water.

Commercial production of titanium has increased enormously since World War II, and the price has steadily dropped. There is every reason to expect that titanium will rank next to iron and aluminum as the most important structural metal of the future. It is used in jet engines; in marine equipment, including propellers and shafts; in machinery; in orthopedic appliances; in sporting equipment, and on and on. It is available in sheets, bars, tubes, rods, wires, sponge, and powder.

When considering the other properties of this wonder metal, however, there are several drawbacks, particularly to the firefighter. Compared with beryllium, titanium receives a clean bill of health as a poison. Its dust is considered only a nuisance. But titanium is one of the metals that burn in massive form.

1. The ignition temperature of a titanium chunk in air is higher than that of magnesium, above 1,300°F (704°C). Large solid pieces do not ignite and burn as readily. Short-time forging and fabricating of titanium alloys in temperatures up to 2,200°F (1,204°C) is possible. There have been some reports of spontaneous ignition of solid pieces of titanium subjected to jets of pure oxygen.

2. Titanium has a higher melting point than magnesium: 3,100°F (1,704°C) as compared to 1,200°F (649°C). The heat of a surrounding fire is unlikely to reach temperatures this high, lowering the chances of creating quantities of molten metal.

3. While the flame temperature of magnesium is close to 2,500°F (1,371°C), titanium burns still hotter. Fine pieces burn slowly, producing great quantities of heat.

4. Strangely, although the solid metal is less combustible than magnesium, the reverse is true when titanium is reduced to a dark gray dust. The ignition temperature of magnesium dust is around 900°F (482°C). A layer of titanium dust can ignite or explode when exposed to temperatures between 720°F and 950°F (382°C and 510°C). Most common ignition sources are at least this hot. When this dust is thrown into the air as a cloud (perhaps by a careless hose stream), an explosion is possible, given an ignition source between 630°F and 1,100°F (332°C and 510°C). One of the reasons for this wide range is that titanium, produced by different processes, varies in its susceptibility to ignition. Titanium powder will decompose water and produce hydrogen. Powders immersed in or wet with water have been known to spontaneously ignite. Titanium

powder ignites in pure carbon dioxide at 1,260°F (682°C) and combines with nitrogen even more eagerly than magnesium at temperatures above 1,475°F (802°C). The powder can also ignite upon impact, because of static electricity or friction, such as is produced by the grinding process. When titanium is being ground, it will produce flying sparks that leave a white trail, ending in several branches. This is unmistakable once seen. (See Table 9–2.)

Most of the recommendations for grinding and machining titanium, and disposal of its sludge, are the same as for magnesium, although a more wary eye must be kept on accumulations of dust. There are no special storage requirements for massive pieces. For finely divided forms, which can also generate hydrogen when moist or oily, recommendations for segregated storage piles, good housekeeping, and covered metal containers are similar to those for magnesium.

Shipping of titanium powder is subject to several DOT regulations regarding containers. Whether wet or dry, the powder must carry a DOT red label: flammable solid. When shipped wet, with 20 percent water, it must be inside tightly closed metal cans not exceeding one gallon (4 liters) each, packed no more than twelve cans to a wooden box. When dry, titanium powder is shipped inside metal containers with soldered or screw caps. These cans must not weigh more than ten pounds (5 kg) each, and the wooden boxes containing them must not exceed a total of 75 pounds (34 kg). (The container must be cushioned by a layer of asbestos wool or rock wool. Dry powder, shipped in one-way metal drums, must also be protected by one of these noncombustible insulating wools.)

As with magnesium, all of the common extinguishers are either ineffective or downright dangerous. Although a coarse spray has been recommended for small quantities of fine pieces, water should be used cautiously. There have been violent reactions when water has been applied to hot or burning titanium. Many of the firefighting methods for magnesium are equally workable for titanium: isolation of burning materials from unburned, metal sandwiches; and the use of such specialized metal extinguishing powders as sodium chloride or graphite. Because the price of titanium is high, chips are salvaged by being dropped into volumes of water. Each pound accumulated is worth a few dollars.

Titanium has already made the "big time" in terms of fires. A 1958 fire involved 2 million pounds of scrap in a pile 300 feet long, 65 feet wide, and 45 feet high. The spread of fire was rapid and intense despite the quick use of hose streams.

ZIRCONIUM AND HAFNIUM

The flammability of a massive piece of metal is not necessarily related to how it will perform as a powder. A chunk of magnesium burns more readily

TABLE 9–2. Ignition Temperatures of Metallic Dusts

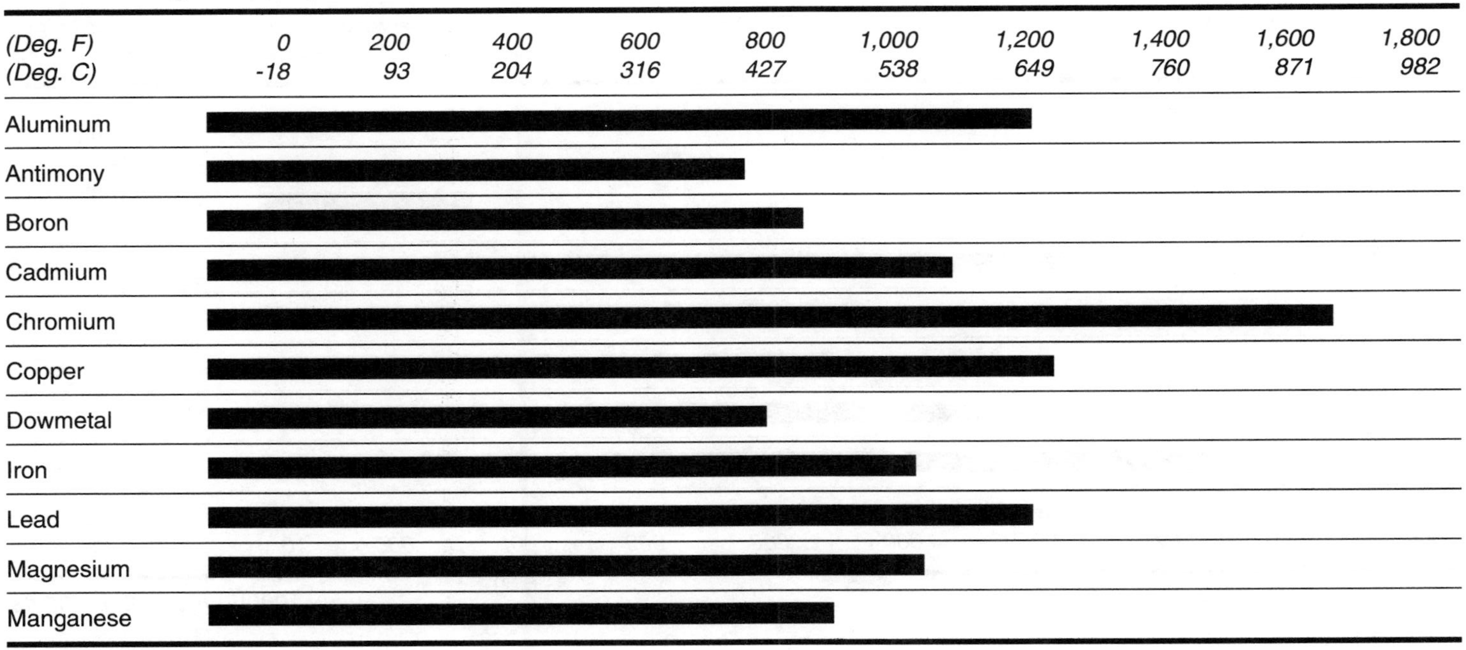

| (Deg. F) | 0 | 200 | 400 | 600 | 800 | 1,000 | 1,200 | 1,400 | 1,600 | 1,800 |
| (Deg. C) | -18 | 93 | 204 | 316 | 427 | 538 | 649 | 760 | 871 | 982 |

Aluminum

Antimony

Boron

Cadmium

Chromium

Copper

Dowmetal

Iron

Lead

Magnesium

Manganese

185

TABLE 9–2. Continued

| (Deg. F) | 0 | 200 | 400 | 600 | 800 | 1,000 | 1,200 | 1,400 | 1,600 | 1,800 |
(Deg. C)	-18	93	204	316	427	538	649	760	871	982
Silicon										
Tin										
Titanium										
Vanadium										
Zinc										
Zirconium as a layer										
Zirconium as a cloud										

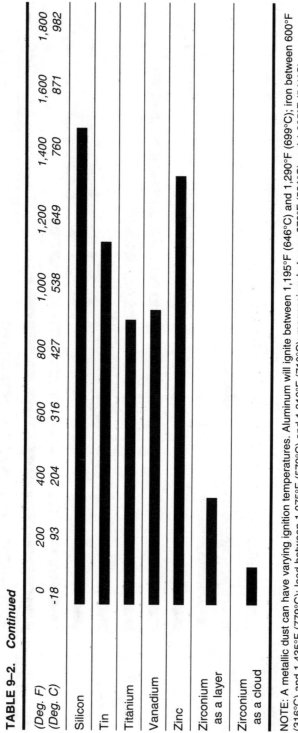

NOTE: A metallic dust can have varying ignition temperatures. Aluminum will ignite between 1,195°F (646°C) and 1,290°F (699°C); iron between 600°F (316°C) and 1,435°F (779°C); lead between 1,075°F (579°C) and 1,310°F (710°C); magnesium between 970°F (521°C) and 1,005°F (541°C).

than a chunk of titanium, while the reverse is true when they are both reduced to dust. **Zirconium** goes a step beyond these two. There are no storage requirements for large pieces of the metal. Its melting point is 3,326°F (1,830°C), and there is little likelihood of the creation of molten metal. Unless somehow subjected to a jet of pure oxygen, massive pieces have not been known to ignite spontaneously. Indeed, a piece of zirconium at least 1/8 of an inch in all directions will not keep burning in air unless assisted by an outside heat source. However, there is an inverse relationship between the size of the individual particles of zirconium and its fire hazard; the danger grows greater as the size of the particles grows smaller. Chips are less dangerous than coarse dust; coarse dust less dangerous than fine dust. Very fine powder is *extremely dangerous*. A DOT red flammable solid label is required.

To evaluate the fire hazard, it is important to know the difference between the different kinds of finely divided zirconium mentioned thus far. The evaluation is done in the laboratory by measuring particle size, using the **micron** (about 1/25,000 of an inch) as a "yardstick." "Coarse" zirconium dust consists of particles measuring about 18 microns; "fine" zirconium powder consists of particles that measure just over three microns. In a fire situation, there is no practical way to tell these forms apart; reliance on labels or correct information given by plant personnel is a firefighter's only source.

Very fine zirconium dust, as a layer or a cloud, will ignite in carbon dioxide at temperatures between 1,150°F and 1,200°F (621°C and 649°C), or in nitrogen, at around 1,450°F (788°C). In air, a layer of dust will ignite at 374°F (190°C). But the most concern comes when this fine dust disperses in air as a cloud. When this happens, the ignition temperature is only 68°F (20°C). This pyrophoricity can be due to frictional heat or generation of static electricity caused by grains of dust bumping against one another. For this reason, zirconium dust must *never* be poured through the air.

Factors that contribute to the flammability of fine zirconium powders include the depth of a pile and the moisture content. Damp powders, containing between 5 percent and 10 percent water, are considered those most likely to explode. They will burn much more vigorously than dry powder. Although hard to ignite, once started, zirconium dust burns more vigorously completely under water than it does in air. Disposal of fine zirconium powder has also proved very hazardous. Extinguishing attempts should be made only with the aid of competent technical advice. Scrap is generally ignited by a long fuse connected to a gasoline-soaked rag with the cans opened by a remote-controlled cleaver, and every precaution is taken against a premature ignition.

The DOT recognizes the importance of particle size in its shipping regulations. Special packaging requirements, including sealed metal or glass containers, cushioning materials, and weight limitations, are required for zirconium powder or sponge, wet or dry. A yellow label, flammable solid, must be attached. However, zirconium powder exceeding 20-mesh* in particle size is not subject to these regulations. (See DOT Hazardous Materials Regulations, Section 173.214, for exact details of regulations.)

The working of zirconium metal is most often done under water or oil-coolant, with the aid of a wet dust collector. Scrap must be cleaned up frequently, and no more than a quart of debris should be allowed to accumulate. The rate of combustion of zirconium chips and turnings varies with the size of the pile, how it is packed, and how wet it is. These fine pieces will not catch fire under water the way the dust will, and they will burn rather peacefully if they are dry and clean. But when wet, they burn with a brilliant white flame, eject burning material, and are likely to burn so fast that no extinguishing method is possible. "Dry powders" such as graphite and sodium chloride will give control over a small fire in wet chips, and they are approved for extinguishment of oily chips. The application of water will intensify the combustion rate, although some tests have shown that, after a brief flare, a small quantity of chips can be extinguished by water. A few chips on fire can be drowned. For powder fires, prevent spread. Ring the fire with an approved metal extinguishing powder. Allow it to burn out. But do not create a cloud of fine dust; it will explode immediately.

Titanium, beryllium, and magnesium have been called the "space age" metals because they are widely used in air- and spacecraft. But this is also the age of the atom, and many metals have come into prominence because of the requirements of an atomic power plant. Zirconium and hafnium are two. Zirconium is unaffected by an atomic particle called a neutron and is used in the inner liners of nuclear reactors. Because the powder is highly flammable, it finds some use in flares, blasting caps, and photoflash bulbs. In addition, alloys of steel and zirconium are bullet-proof.

Zirconium exists as a dark gray powder (loose or pressed into cakes); as a porous mass, called metal sponge; or as a silvery-gray, lustrous, and ductile metal, either in plates, strips, bars, scales, or flakes. Zirconium is resistant to corrosion by acids, alkalis, and seawater. Nitric acid can convert the powder into a shock-sensitive explosive.

Pure zirconium metal is always found in ores with hafnium. Of all the elements, these two are the most inseparable. The fire characteristics, storage, handling, shipping regulations, and firefighting of hafnium are similar to those of zirconium.

Hafnium is 100 times scarcer than zirconium. Hafnium has a great appetite for neutrons. It will slow down the pace of reaction in an atomic pile. The reactor in the nuclear submarine *Nautilus* used hafnium control rods.

---

* This is a common measuring scale for large particles such as sand and gravel. "Mesh" means the size of the holes in the screen used; 20-mesh particles will not pass through a hole 1/20 of an inch in diameter.

ALUMINUM

Aluminum is the most common metal in the Earth's crust and ranks second only to iron in world production and use, well over 9 million tons annually. (An American found a method of producing the pure metal by electrolysis a hundred years ago. It is still the only major structural metal not produced by the smelting of ore.) The metal has many useful properties.

1. Aluminum weighs one-third as much as iron, with a specific gravity of 2.7. Where weight-to-strength ratios are important, such as in utensils and appliances handled by women, or in the frames, engines, and propellers of aircraft, a high percentage of aluminum is still used in spite of the increasing challenge of titanium.
2. Aluminum foils illustrate the metal's malleability; its use in long-distance power lines illustrates its ductility. Pound for pound, aluminum is a better conductor of electricity than copper.
3. Aluminum powder is the same silvery color as the metal, one of the few metal powders not dark gray or black. Mixed with linseed oil, aluminum powder forms the metallic paint so familiar to firefighters.
4. Aluminum reacts rapidly with atmospheric oxygen. Surprisingly, this property is valuable because the metal quickly forms a layer of aluminum oxide that protects it against further oxidation, one reason why aluminum siding is so popular.
5. Aluminum is essentially nontoxic, although the powder can irritate the eyes and respiratory tract.

Properties of aluminum that either limit its usefulness or are potentially hazardous include a melting point of 1,220°F (660°C). The siding on house trailers, for example, can melt so rapidly in a fire that it gives the impression of burning. Creation of molten aluminum is a definite possibility, especially when large pieces are involved. With a melting point this low, structural aluminum is quite likely to fail in a fire situation. When alloyed with other metals to increase its strength at ordinary temperatures, the melting point can be lowered still further. Heavy scrap, or bars, forgings, castings, and sheets are difficult to ignite. In bulk, finely divided aluminum does not ignite easily, but it will burn in air, CO_2, or nitrogen. Powdered aluminum, as a cloud in air, will ignite explosively at temperatures between 1,193°F and 1,292°F (645°C and 700°C). The explosion pressure can be as high as 100 psi (7 atmospheres). No ordinary building can withstand such an explosion without proper venting.

When bulk aluminum powder is involved, our major concern is to prevent the formation of a cloud. Not only will carbon dioxide and dry chemical be ineffective, but the force of their pressurized application will create a cloud. Carbon tetrachloride, applied to aluminum powder, has caused fatal explosions. Water will also stir up the dust and contribute moisture that could intensify the force of the explosion. Our only firefighting alternative is to isolate the fire by the careful application of fine dry sand or approved extinguishing powder. The burning aluminum powder will form a crust and eventually extinguish itself. In the meantime, the area containing the fire should be completely sealed to prevent cloud formation, and all ignition sources should be shut off. (This procedure is also suggested for control of other metal powder fires, such as magnesium.)

When aluminum powder is mixed with a flammable solvent, the use of carbon dioxide may extinguish the fire, although the powder can become exposed to air, form a cake, and re-ignite. Covering with fine dry sand or an approved extinguishing powder is again advisable.

Aluminum powders are used in explosives and in some bronze powders that then share the general hazard. When mixed with iron oxide, aluminum powder forms **thermite**, employed to weld heavy sections of metal. The temperature of burning thermite can reach 5,000°F (2,760°C).

IRON AND STEEL

In spite of increasing use of such metals as aluminum, magnesium, and titanium, iron and steel still form the skeleton of our civilization. Just as compounds are divided into organic and inorganic because of carbon content, metals and metallic alloys are either ferrous or nonferrous. (*Ferrum* is the Latin word for iron; the chemical symbol is Fe.) Most of it is used to make steel. In spite of the furor over steel price increases, it is still the cheapest of metals.

Pure iron is not often found. Iron is a silvery-white, very reactive metal. It will tarnish (oxidize; rust) in air or water. Iron and steel are easily ignited in the form of powders, dusts, or wools. Fourth of July sparklers are often made of iron powders on a metal stick, and they can be lit by a match flame. Unoxidized particles of iron and steel can spontaneously heat. A possible dust explosion hazard exists, since ignition temperatures of powders and dusts are between 600°F and 1,435°F (316°C and 779°C). This wide temperature range reflects the varying compositions due to varying methods of manufacture. When hot, iron and steel dust can react with steam to release hydrogen. To fight a fire in these powders, use an approved metal extinguishing powder.

Some fires have been reported in large piles of steel turnings contaminated by oils. One such fire involved about 1,300 tons of oily steel borings stored on an isolated concrete dock. The pile was taken apart by a "clam-shell" (a type of crane bucket), and the fire extinguished by continuous use of large amounts of carbon dioxide. Water was not used because of the confirmed explosive reaction with the heated metal.

When massive, iron or steel will not burn in ordinary fires; temperatures are simply not high enough. Because these metals melt somewhere around 2,700°F (1,482°C), they do not usually contribute any significant amounts of molten material.

The toxicity of iron powder is minimal, but eye disorders can result if particles are not removed promptly, and fumes from heated iron can cause some lung trouble. Nevertheless, iron has nothing like the toxic quality of beryllium and some other metals.

Steel-Making

Without wandering too far afield, some idea of how a basic industry like this operates is valuable.

Iron ore is heated in blast furnaces. The removed impurities are called **slag**. The remaining metal, 90 percent pure, is known as **pig iron**. Depending upon the way pig iron is cooled, suddenly or slowly, it becomes one or another of the cast irons. Some cast irons can withstand high temperatures without noticeable distortion. Others will break if suddenly heated and cooled. (This can happen when hose streams strike a cast-iron column during a fire. Structural steels must be used for construction.) If the pig iron is remelted and further purified in a puddling furnace, it becomes **wrought iron**, softer but tougher than cast iron. The major difference between these two irons is in their carbon content. Wrought iron has less than 1.0 percent carbon, pig or cast iron from 2.0 to 5.0 percent. In between these percentages are the carbon steels, deliberately created according to specifications in open hearth, Bessemer, or electric furnaces.

Iron rusts; so does steel. Unlike aluminum oxide, which hangs on tight and prevents further atmospheric assaults, iron oxide continually flakes off and exposes a new surface to fresh oxidation. In time, a large piece of iron or steel will rust away.

STEEL ALLOYING METALS

Special steels are made by alloying iron with other metals. Many of these metals find their principal industrial use (over 100 million tons each year) in the formation of these alloy steels. Some of these metals and their properties appear in alphabetical order below.

Chromium

Chromium, alloyed with iron and nickel, forms stainless steel, which is not only rust resistant but also has great strength. Chromium is a silvery-white, hard, and brittle metal that will take a brilliant polish. Automobile chrome is steel with a thin layer of nickel covered by chromium. Although chromium does not tarnish in air and resists oxidation, it burns in oxygen at high temperatures and decomposes water when red-hot. It melts at 2,929°F (1,609°C). It is only a moderate dust explosion hazard, since the ignition temperature of a cloud under favoring conditions is about 1,650°F (899°C). Chromium will produce hydrogen gas from strong acids. It is essentially nontoxic. A strategic metal that must be imported into the United States, it is available in several forms.

Cobalt

Cobalt is alloyed with iron to make permanent magnets. It is a steel-gray shining metal that is stable in air or water at ordinary temperatures and has a low order of toxicity. Annual world production is about 20,000 tons.

Manganese

Ninety percent of the **manganese** used by industry goes into the alloying of other metals. Manganese steels are tough and resist wear. They are used in safes, steam shovel teeth, and rock crushers. The pure metal is a silvery to reddish-gray color. It melts at 2,246°F (1,230°C). Manganese dust will ignite in air at 840°F (449°C), in pure nitrogen above 2,000°F (1,093°C), and burns with an intense white light. It will produce hydrogen in contact with strong acids, or when it reacts with water or steam at temperatures of 200°F (93°C). Chronic manganese poisoning and paralysis can result from continued inhalation of fumes or dusts. No DOT label is required.

Molybdenum

Molybdenum steels have high tensile strength and heat resistance, ideal for high-speed tools. Molybdenum is a silvery-white metal. The powder will react with steam at 1,300°F (704°C) and is a slight dust explosion hazard.

Nickel

Nickel steels have great hardness and high tensile strength. They can be used in automotive parts or even armor plate. Nickel itself is a hard, silvery metal. It is not considered toxic or a serious fire hazard, although a dust explosion is possible. The annual world production of this widely used metal is well over 475,000 tons.

Silicon

Silicon steels are used in springs because of their flexibility. Silicon is a nonmetallic element, commonly found in two forms: a dark-brown amorphous powder that will ignite in air at 1,425°F (774°C) and silvery leaflets that will not burn.

Tantalum

Tantalum steels, used in tools, are very hard. At higher temperatures, the bluish metal or black powder becomes reactive. Although it has a low order of toxicity, it has caused some skin injuries.

Tellurium

A nonmetallic element, **tellurium** is used in iron and stainless steel castings. The dark gray crystals or amorphous powder will melt at 846°F (452°C) and burn in air with a greenish-blue flame. There is a moderate dust explosion hazard. Probably toxic, tellurium can be absorbed through the skin. People who work around this element often have a breath that smells like garlic.

Tin

Tin cans are sheet steel covered by a thin layer of this silvery-white and ductile metal with a melting point of 450°F (232°C). In powder form, tin will oxidize in wet air. The dust cloud ignites at 1,165°F (629°C). A solid piece of tin will decompose carbon dioxide at 1,020°F (549°C), react with water vapor at 1,200°F (649°C), and ignite and burn with a luminous white flame at 2,370°F (1,299°C).

Tungsten

Tungsten steels are used in high-speed tools because they retain their hardness when heated. Tungsten (wolfram) is a steel-gray metal with a high luster. The finely divided powder can be pyrophoric, and the metal reacts violently with various oxidizing agents.

Vanadium

Vanadium steels, used in axles and gears, are shock resistant. Vanadium is a steel-gray, extremely hard metal. The dust will ignite in air at 932°F (500°C). Prolonged exposure to concentrations of dust can lead to lung disorders.

ZINC

Besides alloying and painting, "galvanizing," which gives metal a protective coating of **zinc**, is another way to protect iron from oxidizing. Like aluminum, zinc protects itself with an oxide layer against atmospheric oxygen. A little less than half of the annual world production of 4.5 million tons of zinc is used for this purpose. Zinc is alloyed with copper to form brass and with copper and nickel to form German silver. It is also used in metal spraying.

Zinc is often called **spelter** in the metal trades. It is a bluish-white, malleable, and ductile metal and a bluish-gray powder. It is available in forms ranging from slabs to dust. In dust form, it is shipped in cartons, boxes, barrels, and drums.

Hazards of Zinc

The flammability hazard of zinc is a variation of a familiar story. Massive forms will not ignite easily, but they burn vigorously in air with a bluish-green flame, giving off quantities of white or bluish smoke once they catch fire. The metal melts at 786°F (419°C), a temperature low enough to create molten zinc readily in a fire. The dust forms explosive mixtures in air, with clouds igniting at 1,110°F (599°C). Bulk dust, *when wet,* or when contacted by acids or lyes, will liberate hydrogen and may heat spontaneously to ignition. Powders must be stored in cool, dry areas, well ventilated, and separated from acids, lyes, and moisture.

A fire in zinc powder should be smothered with an approved metal extinguishing powder. The use of water on massive zinc pieces is subject to all the hazards noted for other metals with low melting points, plus one addition: Water, carbon dioxide, and halogenated hydrocarbons can react vigorously with burning zinc to form toxic gases. While the metal itself is not toxic, when heated or burning, zinc forms oxide fumes that are poisonous and cause "fume fever" or "brass chills," characterized by a sweet taste, dryness in the throat, aches, chills, fever, and nausea. Firefighters should wear self-contained breathing apparatus to guard against these fumes.

TOXIC METALS

Besides beryllium and zinc, several industrially important metals add toxicity to their flammable hazard. The first of these is antimony.

Antimony

Antimony (stibium) is silvery-white, hard, and brittle with a melting point of 1,166°F (630°C). Although the pure element has no commercial use, antimony alloys and compounds are employed to harden lead in storage batteries and can be found in Babbitt metal (5 to 18 percent), in solder, printer's type, transistors, paints, ammunition, and glass. It is shipped as 55-pound (25-kg) bars or pigs, as lumps, or in finely divided form. Antimony does not react with air at room temperatures but, when heated above 780°F (416°C), it ignites and burns with a bright bluish flame. When molten, it reacts with water to produce hydrogen gas and then reacts with the hydrogen to form an extremely toxic gas called **stibine**. This gas is sometimes given off by lead storage batteries. When heated, antimony can emit toxic fumes.

Bismuth

Bismuth is hard and brittle, grayish-white with a reddish tinge, and has a bright luster. It is available in various forms. It burns in air with a blue flame. Although considered one of the least toxic of the heavy metals, it can cause kidney damage if taken into the body and should be considered a poison. Its melting point is 520°F (271°C). When 50 percent bismuth is combined with various percentages of lead, tin, and cadmium, it forms a fusible alloy that will melt at a specific temperature. These alloys form fusible links that open sprinkler heads and close fire doors.

Cadmium

Cadmium is a grayish-white powder or a bluish-white, ductile, malleable metal soft enough to be cut with a knife. The metal tarnishes in moist air and burns when heated. Its melting point is 610°F (321°C). Cadmium has a wide range of uses. Cadmium plating of containers has caused food poisoning. *When heated, it emits highly toxic fumes.* A brief exposure to high concentrations can cause death from pulmonary swelling.

Lead

Lead is a familiar, dull gray, very soft, heavy, highly malleable, and ductile metal. Over 3 million tons are produced annually. It is used for such purposes as cable coverings, plumbing, ammunition, radiation shields, glass, printer's type, and storage batteries. Its melting point is 621°F (327°C). At high temperatures, it burns with a white flame. When it is at red-heat, rapid oxidation can be caused by air. The dust explosion hazard is slight; a cloud can explode between 1,075°F and 1,310°F (579°C and 710°C). Lead poisoning is one of the most common of occupational diseases. It is a cumulative poison, which means that a little bit at a time can add up to a toxic dose because the body does not get rid of it easily. Poisoning can come from inhalation of dusts, fumes, or vapors, or sometimes through the skin. When it is inhaled, the symptoms develop quickly: anemia, pains in the joints and muscles, headaches, dizziness, and abdominal tenderness. If lead dust or heated lead is involved in a fire, firefighters should wear protective equipment and self-contained breathing apparatus.

Mercury

Mercury is the only metal that is a liquid at ordinary temperatures. It is a shiny, silvery, surprisingly heavy liquid. A 16-pound iron shotput or a lead ball will float on a pool of quicksilver. Shipping bottles of one, five, or ten pounds are quite small. Production is measured in 76-pound flasks, about 270,000 of which are produced annually. Mercury is used to extract gold and silver, in arc lamps, in explosives (fulminate of mercury), in thermometers, and, most commonly, in silent mercury switches.

Mercury melts at -38°F (-39°C) and boils at 676°F (358°C). These temperatures are important because mercury is a cumulative poison that can be taken into the body by inhalation, ingestion, or even through unbroken skin. Air saturated by mercury vapor at 68°F (20°C) contains more than 100 times the toxic dose. Danger becomes even greater at higher temperatures as the boiling point is approached. Firefighters must be careful to protect themselves against these vapors. Symptoms of mercury poisoning include excess salivation, pain upon chewing, and nervous and mental disorders.

Thallium

Thallium is a bluish-gray metal that resembles lead. It is also toxic and should be carefully handled. Skin contact can be dangerous. One thallium compound is widely used as a rat poison. It decomposes water when red hot, releasing hydrogen.

GLAMOUR METALS

The remaining metals are the glamorous members of the tribe, the precious metals used in jewelry or those so rare there is only the slightest chance of meeting one in a fire situation. Still, strange things happen, and if a fire involves a precious metal, some of us are likely to be there.

Copper

The only inexpensive member of this group is **copper**, available in many forms and shapes. Copper is malleable and is one of the two metals that have a definite color. Hammered copper has been an ornamental metal throughout history. Because it is also ductile and an excellent conductor of electricity, much of the copper produced today goes into wire. For a long time, copper was second only to iron in industrial importance. Since the war, aluminum has taken over the second spot, although 6 million tons of copper are used annually for ornaments, wires, machinery, piping, cooking utensils, and in alloys. Brass is copper (65 to 80 percent) and zinc. Bronze is copper (80 to 95 percent) and tin. Monel metal is two-thirds nickel and one-third copper. Where nonsparking materials are necessary, such as around flammable liquids, an alloy of copper and beryllium is often used.

Copper melts at 981°F (527°C). A dust cloud can ignite under favorable conditions at 1,290°F (699°C). Although the pure element is low in toxicity, many copper compounds are poisonous.

Platinum Metals

Osmium is one of the platinum metals. It is white with a bluish cast, hard, and the heaviest of all elements. The dust will slowly oxidize at room temperatures. It is somewhat sparingly used in pen points or as a hardener for platinum. Although the metal is not highly toxic, upon heating in air it gives off a pungent, nauseating, poisonous fume, osmium tetraoxide. Some of the other platinum metals have poisonous possibilities. All of them are expensive. Platinum itself has an irritating dust that may cause skin disorders.

Ruthenium dust ignites above 1,300°F (704°C) and emits toxic fumes when heated. The other three platinum metals—**rhodium, palladium, and iridium**—as far as is known, have a low order of toxicity and slight fire hazards.

Rare Earths

One group of fifteen elements is called the "rare earths." Actually, they are not earths, but metals, and in a few cases, not particularly rare. However, most of them are uncommon and require only a brief mention.

Cerium is the most abundant of these elements, more common than tin, silver, or mercury. It is a steel-gray, ductile, malleable metal, soft enough to be cut by a knife. This test is not advisable, because the pure metal has been known to catch fire when merely scratched. It will ignite above 300°F (149°C) and burn with a bright white flame. As with all metals, finely divided particles increases the fire hazard. It melts at 1,459°F (793°C). When heated, cerium will decompose water and liberate hydrogen gas. It is not considered highly toxic, although, like the other rare earths, little is really known about its toxicity. Misch metal, used in alloys and cigarette lighter flints, is made from a combination of rare earth metals, with over half of the alloy being cerium.

The other metals in this group are:

Dysprosium	Gadolinium	Lutetium	Samarium	Ytterbium
Erbium	Helmium	Neodymium	Terbium	Promethium
Europium	Lanthanum	Praseodymium	Thulium	

Most of these metals have properties similar to those of cerium. They will ignite in air between 300°F and 350°F (149°C and 177°C) (with the exception of lanthanum, known to ignite at 824°F or 440°C). They have varying rates of reaction with air and water. Many of them are stored in light mineral oil or sealed plastic. Two other metals, **yttrium** and **scandium**, have closely allied properties. They are very expensive and rare, hardly out of the lab yet. The first pound of scandium was produced in 1960.

Remaining Metals

Eight other elements, many of them unfamiliar, complete the metals roster.

Boron is a black crystalline powder or soft brown amorphous powder that will ignite in air. It may cause poisoning if swallowed or inhaled. It is used to make Pyrex glass.

Gallium is a silvery metal that melts at 85°F (29°C) and boils at 3,601°F (1,983°C). The metal expands when it solidifies and can break glass containers. It is believed to have a low amount of toxicity.

Indium, a silvery-white metal, softer than lead, can be toxic if taken into the body. It melts at 313°F (156°C) and burns in air with a blue flame.

Silver is the best metallic conductor of heat and electricity. Avoid inhaling dust. It resists oxidation but reacts with sulfur.

Gold is a shiny, yellow metal, not toxic or acted upon by air or water.

Germanium, a grayish-white metalloid, will burn in chlorine and bromine. It has a low order of toxicity, although some poisonings have occurred.

Rhenium, a silvery metal or gray-to-black powder, is stable in air and very scarce. World supply is estimated at only 100 tons.

Niobium, a silver-gray metal, reacts to oxygen and halogens only when heated and starts to oxidize at 392°F (200°C). Toxicity is unknown.

SUMMARY

Combustible metals are extremely hazardous to firefighters in that some react with water, and the majority react violently when water is used to extinguish a fire in which they are involved. Since water is the most acceptable extinguishing agent and easiest to apply, firefighters have a tendency to get into trouble before assessment of the consequences can be determined. Combustible metals burn so hot that they are capable of liberating oxygen and hydrogen from the decomposition of water. In late 1989 and early 1990, firefighters in the Seattle area experienced fires in buildings that were apparently set using flammable solids as the accelerant. These types of fires are almost impossible to extinguish. Tests show that flashover in a large commercial building could occur within three minutes from ignition. Couple this with the acceleration caused by the application of water to a flammable solid and the results are devastating. In September of 1989 such a situation claimed the life of a Seattle fire captain. The firefighter who entered the building with the captain was able to get out of the building. However, he was suffering from extreme heat exhaustion with a core body temperature measured at 105° Fahrenheit because of exposure to the high heat generated by the burning of flammable solids. Signs of the use of a flammable solid accelerant are hot white fire, arcing flames, extreme high interior temperatures, and increased fire intensity early in the incident. This is a no-fight situation. Protect exposures.

REVIEW QUESTIONS

1. Name five properties shared by most metals.
2. Discuss the importance of the size and shape of a metal in determining its degree of combustibility.

3. What is the meaning of the word pyrophoric?
4. Give two reasons why the use of water can be dangerous around metal fires.
5. Discuss some of the sources of oxygen for burning magnesium.
6. What are the comparative merits and drawbacks of fog patterns and straight streams on magnesium fires?
7. Why are automatic sprinklers recommended for magnesium storage?
8. Describe two extinguishing powders that are recommended for some metal fires.
9. Name five metals that have a toxic hazard.
10. How do massive titanium and finely divided titanium compare in hazard with magnesium?
11. Discuss the hazards of zirconium powder.
12. What is our major concern in a fire involving aluminum powder?

BIBLIOGRAPHY

1. National Fire Protection Association, 1 Batterymarch Park, Quincy, MA.
   Standard No. 10, "Portable Fire Extinguishers."
   Standard No. 65, "Code for the Processing and Finishing of Aluminum."
   Standard No. 480, "Standard for the Storage, Handling, and Processing of Magnesium."
   Standard No. 481, "Standard for the Production, Processing, Handling, and Storage of Titanium."
   Standard No. 482, "Guide for Fire and Explosion Prevention in Plants Producing and Handling Zirconium."
   Standard No. 651, "Code for the Prevention of Dust Explosions in the Manufacture of Aluminum Powder."
2. "Pyrophoric Metals, a Technical Mystery," *NFPA Quarterly*, October 1957.
3. "Combustible Metal Fires," *Fireman's Magazine*, May 1958.

FURTHER READINGS

Edwards, R., and D. Edwards, *Fire Chem II, S.A.F.E. Films Inc.*, 3326 Bentwood S.E., Grand Rapids, MI, 1989, pp. 113–128. Text supported with videotapes.

VISUAL AIDS

"Magnesium Fire, City of Industry," American Heat Video Production, 240 Sovereign Ct. Ste. C, St. Louis, MO, Vol. 5, Program 7.

"Flammable Solids as an Accelerant in Arson Fires," American Heat Video Production, 240 Sovereign Ct. Ste. C, St. Louis, MO, Vol. 4, Program 12.

Plastics

Not satisfied with flammable liquids and gases or combustible solids, metallic and otherwise, American industry is busily creating and using an enormous group of synthetic materials that can also catch fire: the plastics. Defined in simplest terms, a plastic is a material that becomes moldable at least once because of chemical treatment, heat, or pressure and that subsequently hardens into a new shape.

It all began in 1833 when a French scientist discovered that if he treated cellulose, the basic component of plants, with nitric acid, he obtained a substance that burned cleanly and very rapidly. He called it **xyloidin**. At that time, military gunpowders produced a puff of smoke that gave away soldiers' positions. If xyloidin could be turned into a smokeless powder, it would be invaluable. It immediately became one of the first scientific babies to be drafted, but it proved difficult to train. Explosions occurred with distressing frequency as experimenters sought to apply xyloidin to military use. There was a noticeable lack of enthusiasm for continued research.

So matters stood for about 30 years, until concern began to be expressed about the fate of elephant herds being slaughtered for their ivory. A billiard ball manufacturer offered a $10,000 reward to anyone who could produce a substitute that would meet the exacting requirements of a cue ball. Tinkering with xyloidin, an American printer named Hyatt added camphor, which stabilized it enough to be worked. For awhile, from Jumbo's viewpoint, it seemed that xyloidin had left the battlefield for the pool hall in the nick of time. Alas, the new material proved too brittle for the fifteen numbered balls. But Hyatt had created the first usable plastic. In the years since its discovery, this mixture of xyloidin and a camphor **plasticizer*** came to be known by its brand name, **Celluliud**.

*A plasticizer is any material added to improve the flexibility and other properties of a basic substance like xyloidin.

Celluloid proved to have many excellent and economically useful properties. It was, it seemed, good for a lot of purposes outside the billiard parlor. It was resilient and tough, resistant to water and oils, could be produced in many colors and shapes, or buffed and polished to a high, attractive luster. Most important, it was cheap and easy to make. A whole generation of young men nearly choked to death in their celluloid collars. More and more uses were found for it: dental plates, piano keys, X-ray and photographic films, Ping-Pong balls, buttons and ornaments, shoe heels, toilet articles of various kinds, fountain pens, spectacle frames, toys, lacquers, paints, and patent leather. But as more was learned about its properties, Hyatt's invention came to be known by another name, **pyroxylin**, an appropriate word meaning "fire wood." One disadvantage of this new material was a degree of fire hazard still unmatched by any other plastic. A closer look at its beginnings will explain why.

CELLULOSE NITRATE

The basic raw material, xyloidin, is known today as cellulose nitrate or nitrocellulose. It is produced by dipping some form of cellulose, generally wood pulp or cotton linters (small tufts of cotton stuck to the cottonseeds after ginning) into a solution of concentrated nitric and sulfuric acids. The process is called **nitration**. Any excess acid is removed by washing and boiling the fibers, which are then pulped before a final washing. The result is a white or near-white cottony fiber, chips, or powder. By varying the strength of the acids, the dip time, or the proportion of cellulose to acid, one obtains materials of differing properties. What nitration accomplishes is to attach nitrogen and free oxygen (ONO_2) to the cellulose molecule. The solubility, the uses, and even the names of the various forms of cellulose nitrate depend upon the percentage of nitrogen they contain (see Table 10–1). More significant to the fire service is that the higher this nitrogen percentage, the more potentially dangerous the resulting material.

Cellulose Nitrate Fire Hazards

The chemical formula for one form of cellulose nitrate is $C_{12}H_{17}$ $(ONO_2)_3O_7$. Within the parentheses are three nitrogen atoms and free, available oxygen. Another form of nitrocellulose has four such groups attached. The more oxygen in its makeup, the merrier any form of cellulose nitrate burns, even in the absence of air. Once ignited, the combustion process is very thorough. It cannot smolder, for there are no places where oxygen cannot reach, since it is built in. Nitrocellulose is all surface. Methods of extinguishment based upon the exclusion of air are just a waste of time.

TABLE 10–1. Types of Nitrocellulose

Percentage of Nitrogen	Names	Solubility	Uses	DOT Labeling
10.9–11.2	Pyroxylin Plastic, soluble Nitrocellulose	Alcohols, esters, ketones	Paper coatings, plastics, low-odor lacquers, printing inks	Flammable solid
11.3–11.7	Pyroxylin Plastic, soluble Nitrocellulose	Alcohols ether-alcohols, esters, ketones	Cellophane and paper coatings, alcohol-soluble lacquers, textile coatings	Flammable solid
11.8–12.2	Pyroxylin Plastic, soluble Nitrocellulose Collodion Photocotton	Ether-alcohols, esters, ketones, glycols	Dopes, adhesives, coatings artificial leather, collodion, fast drying lacquers	Flammable solid
12.6–12.8	Pyrocellulose Pyrocollodion	Acetone	Propellants	Explosive A
13.0–13.8	Guncotton (also called smokeless powder)	Acetone	Propellants, explosives, smokeless powder	Explosive A

Pound for pound, cellulose nitrate does not produce as much heat as burning wood or paper. But the comparison is both misleading and meaningless, for it burns more than a dozen times faster. A ton of nitrocellulose can be completely consumed in a little over a minute and a half. The liberated heat is simply enormous, with flames extending like white blowtorches in all directions. Spectators hundreds of feet away from a nitrocellulose fire have been injured. The quickness of the spread of such a fire is awesome and extremely dangerous. Because of burning rate alone, cellulose nitrate would qualify handily on most lists of extremely hazardous materials. Yet there is more trouble locked inside this molecule; nitration has made it unstable.

Like acetylene, cellulose nitrate will decompose, break apart, and release energy. But, while acetylene was primarily sensitive to pressures above 15 psig, nitrocellulose is sensitive to heat. Temperatures above 300°F (149°C) almost invariably cause decomposition; temperatures as low as 100°F (38°C) begin it. Such seemingly innocent sources of heat as steam pipes, friction, electric light bulbs, or even sunlight shining through unpainted windows are enough. Not being dependent upon the presence of air, decomposition is extraordinarily difficult to prevent. Once it starts, it is self-sustaining and constantly growing, heat-producing yet flameless. Soon the signs begin to appear—an acrid odor, a brownish discoloration,

PHOTO 10–1. DOT explosive label. The labels for class A, B, and C explosives are orange with black lettering and designated as Explosive A, Explosive B, or Explosive C.

blistering, quantities of brownish-to-white gases. All the while the heat continues to build until, finally, the ignition temperature of nitrocellulose is reached. This is how most fires in cellulose nitrate storage are caused and one of the reasons why the addition of camphor is so important: It inhibits (discourages) the decomposition process. But aged stock can lose camphor through the normal evaporation of this volatile material. When it leaves, the remaining plastic becomes extremely vulnerable.

Even the heat of decomposition does not complete the list of hazards. Decomposing cellulose nitrate produces gases in quantity. A single pound yields three cubic feet (.075 cubic meters) of gases, enough to make 400 cubic feet (10 cubic meters) of air explosive. Quantity storage can create enough gas to destroy walls, or even buildings, by the force of pressure alone. These gases are not only flammable and capable of exploding before or during a fire, but they are also highly toxic. Three of them of major concern to firefighters are listed in Table 10–2. The flammability of these gases varies, as does their water solubility, which measures fire-fighters' ability—or a sprinkler system's—to wash them out of the air. However, all three are poisonous. The presence of the nitrogen oxides is a final gift from the nitration process. These brownish-to-tan gases are insidious. They have little irritating effect on the respiratory passages, but once they hit the moisture of the lungs, they turn into nitric acid. Toxic, even fatal, symptoms can be delayed for several days. These gases are produced not only during decomposition, but also, in somewhat smaller amounts, when cellulose nitrate is burning.

TABLE 10–2.   Gases Formed by Cellulose Nitrate

Gas	Flammable?	Toxic?	Water Soluble?
Carbon monoxide	Yes	Highly	No
Hydrogen cyanide	Yes	Highly	Yes
Nitrogen oxides	No	Highly	No

Shipping and Storage

The regulations regarding the transportation of cellulose nitrate depend upon the answers to many questions, among them:

1. How high is the percentage of nitrogen?
2. What is the physical form of the nitrate?
3. Has the nitrocellulose been dissolved, or is it moist or even dry? If it is moist, what kind of damping medium has been used?

A plasticizer, most often camphor, protects celluloid, although dry, from decomposing. Celluloid can be found in many sizes and shapes: sheets, rolls, tubes, rods, or as the finished article, a toy, an eyeglass frame, a toilet seat. Most manufactured articles are not under DOT shipping regulations. Scrap celluloid, often susceptible to decomposition because of contamination, may also be shipped unless there are signs of decomposing. Then it must be under water. (The regulations for nitrate photographic and X-ray film will be considered later in this chapter when it is compared with Safety Base, cellulose acetate. Note: Safety Base cellulose film has been in production since 1952; therefore, encounters with cellulose nitrate film in emergencies has become rare. However, it is still out there.)

Although the hazards of celluloid are extreme, cellulose nitrates with still higher percentages of nitrogen have also found widespread uses. Since the initial series of explosions that so discouraged the military, improved techniques have been found to tame xyloidin. Through the use of solvents, it can be softened into a dough that dries to a brittle solid suitable for making smokeless powder. Dry or even moist guncotton or smokeless powder can be exploded by a detonator. It was used as a high explosive for a time but has been replaced by less treacherous materials such as dynamite and TNT. The high nitrogen forms of cellulose nitrates are considered to be explosives and subject to stringent shipping regulations.

When nitrocellulose is totally dissolved in one of the suitable solvents mentioned in Table 10–1, it is considered "only" as dangerous as the solvent itself. For example, when nitrocellulose is dissolved in a mixture of 70 percent ether and 24 percent alcohol, it forms **collodion**, a pale yellow syrupy liquid with an ether odor. Exposed to air in a thin layer, collodion gives a tough colorless film suitable for some medical and industrial uses. Collodion is shipped in glass bottles, in 1- to 10–pound cans (4.5 kg), and in 30- to 60-pound drums (14 to 28 kg) under a DOT red label: flammable liquid. All of these containers should be kept tightly closed when not in use. Collodion has a flash point below zero.

Whatever it may be called, industrial nitrocellulose with a lower percentage of nitrogen is generally shipped in steel 55-gallon (207-liter) drums, wet or dampened (not dissolved) with some sort of liquid, most often butyl alcohol, denatured ethyl alcohol, isopropyl alcohol, or water. When uniformly wet with 20 pounds (9 kg) of water to 80 pounds (36 kg) of dry material, cellulose nitrate carries a DOT red label: flammable solid. However, often the final use of nitrocellulose does not permit the use of water as a damping medium. When shipped dissolved in one of the alcohols, the percentage of liquid required depends upon the physical form of the nitrate. If it is fibrous or cottony, at least 30 percent, by weight, of alcohol or solvent is necessary. When granular or flaked, it must contain 20 percent by weight. The flash point of the liquid used must not be lower than 30°F. (-1°C.). In all cases, the drum carries a DOT red label: flammable liquid. There are also special limitations on the amounts that can be shipped by rail express.

Naturally, if the protection of the damping medium is lost, either through leakage due to physical damage to the drums or because the drums are not periodically inverted to assure an even distribution of moisture, the nitrocellulose will dry out. When this happens, the hazards are greatly increased. The presence of a flammable liquid also creates additional problems.

Storage regulations for cellulose nitrate are based on two underlying principles. The first is fire prevention in grim earnest. All sources of heat, even those lukewarm, must be rigidly controlled. Water should be used as a coolant on machines, wire guards placed around common ignition sources, sparks kept to a minimum, stock protected against frictional heat, and an absolute prohibition enforced against smoking and matches. There should be periodic inspections for signs of decomposition, to make sure that drums have not been physically damaged and that they are being inverted every 30 days. Of course, good housekeeping and all it entails is a necessity. Behind these fire prevention activities (and those mentioned in the standards listed at the conclusion of this chapter) must be knowledge of where this dangerous material is stored. Whether in film projection booths, X-ray vaults in hospitals, industrial establishments where paints, lacquers, or organic coatings are manufactured, or where articles of pyroxylin are stored or machined, a nitrocellulose fire should never come as a surprise. Fire departments should know where the potential exists.

The second principle: A large amount of material should not be allowed to decompose all at once. If a fire begins, it should involve only a relatively small amount. To this end, nitrocellulose should be segregated in storage. It should be stored in special isolated buildings, in vaults or cabinets, in work areas separated from one another by floor-to-ceiling partitions, or in special stock rooms of a fire-resistive construction. Even the use of tote boxes, wrappings, cartons, or paper envelopes to contain small amounts is helpful. It is curious, but logical once it is considered, that when cellulose nitrate has to burn *through* something, even a paper envelope, fire spread is greatly slowed.

This principle of segregation should be backed up by a building protected by decomposition and explosion venting and—very important—by a sprinkler system. Although sprinklers may not operate immediately because of the low initial temperatures of decomposition and although they may be overpowered by the resulting fire, nothing can replace a system with closely set heads, fed by a strong water supply.

Fire Fighting of Cellulose Nitrate

The precautions necessary to prevent and lessen the consequences of a cellulose nitrate fire have been discussed. The following is a summary of what such a fire can be like:

1. Burning or decomposing nitrocellulose produces toxic gases in great quantities. Even in a fire involving a small amount of pyroxylin, firefighters have been hurt by these gases. Whenever cellulose nitrate is on fire or decomposing, firefighters *must* wear self-contained breathing apparatus. Every effort should be made to keep them upwind of a large fire. The area surrounding the fire should be evacuated.
2. Explosions can be caused in several ways:
    a. The amount of gas produced by burning or decomposition can cause enough pressure to force out walls.
    b. These gases are flammable. Produced in such quantity, they may pass quickly above the upper flammable limit when close to their point of origin. They will ignite or explode when they find an oxygen source (places where firefighters are likely to be): at windows, in doorways, in adjacent rooms.
    c. Flammable liquids are often associated with nitrocellulose storage. Not only can their vapors ignite, but if a drum of nitrocellulose catches fire, the intense heat it will generate can cause a vapor pressure explosion in nearby drums, no matter what liquid is inside them, even water. A warehouse full of closely stored containers can become a gigantic, murderous, popcorn popper.

There is only one extinguishing method: An ocean of water, quickly applied to the entire surface, is the only method with a chance of succeeding. Large amounts of water to combat the heat of combustion, quickly applied to combat the rapidity of the spread, is a firefighter's only chance. The only firefighting appliance in a position to do this immediately is an automatic sprinkler system, if explosions have left it intact.

Even when the fire is finally extinguished, the problems may not be over. There remains overhaul and possible disposal. Overhaul procedures must be carefully planned. Scrap pyroxylin after a fire may be extremely prone to decomposition after heating and possible contamination. If it is to be disposed of, it should be burned in the open, in small amounts, away from buildings, with due consideration of what lies downwind.

The unwelcome properties of cellulose nitrate, the fact that it discolors upon aging or in the sunlight, that it will dissolve in many common solvents, that it withstands heat poorly, and, above all, its hazards, led to a search for other plastic materials without these drawbacks. Nitrocellulose was the pioneer. It proved that there was a ready market for synthetic materials and showed the technology needed to create them. How well industry succeeded in finding substitutes for pyroxylin is the subject of the rest of this chapter.

HOW PLASTICS ARE MADE

The worst came first. Compared with those of cellulose nitrate, the hazards of other plastics seem minor. In most cases, plastics burn no faster than wood or paper. Their storage requirements, transportation regulations, and methods of extinguishment are similar to those of ordinary Class "A" materials. Nevertheless, they do have a few distinctive dangers of their own: the toxicity of their smoke, their methods of burning, and the hazardous company they keep.

Firefighters will meet plastics not only in places where they are manufactured, fabricated, or stored, but increasingly in every structure fire. Hundreds of pounds are in every modern home: in appliances, insulation, or even as part of the building itself.

Seemingly, there is a plastic for every use. What does your business need? A rigid or flexible container? A special piece of intricate shape? A foam? A film? A fiber? A coating? A plastic is either ready to fill the bill immediately, or, if you wait a minute, one will be tailor-made to meet your specifications as to resistance to cold, heat, and chemicals, weight, strength, hardness, color, electrical conductivity, or whatever else you require.

TAILOR-MADE PLASTICS

Polyethylene

How to build a plastic to order? It all begins with a basic material like our old acquaintance, ethylene gas. As you remember, ethylene has a double bond between its carbon atoms (see Figure 10–1).

In a sense, a double bond is under some stress. If prodded, the carbon atoms will revert to single bondage. In the case of ethylene, this helping

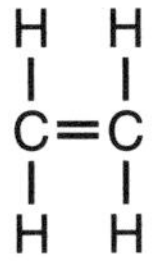

FIGURE 10–1. Ethylene (C_2H_4).

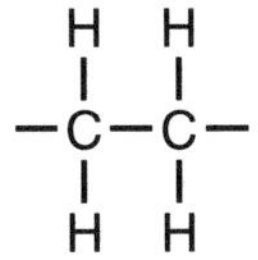

```
 H   H
 |   |
-C - C-
 |   |
 H   H
```

FIGURE 10–2. Rearranged ethylene molecule, the ethyl radical.

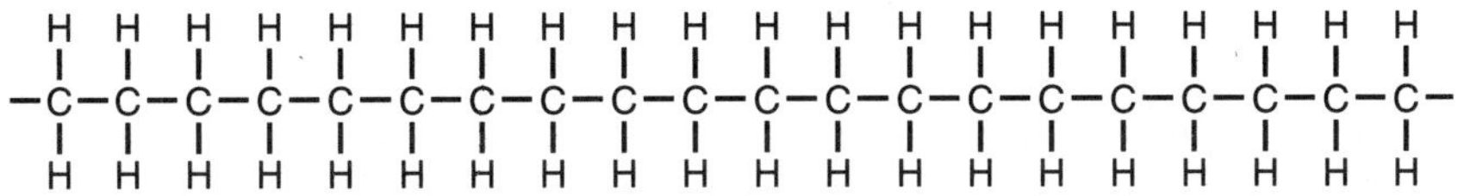

```
 H  H  H  H  H  H  H  H  H  H  H  H  H  H  H  H  H  H  H  H
 |  |  |  |  |  |  |  |  |  |  |  |  |  |  |  |  |  |  |  |
-C--C--C--C--C--C--C--C--C--C--C--C--C--C--C--C--C--C--C--C-
 |  |  |  |  |  |  |  |  |  |  |  |  |  |  |  |  |  |  |  |
 H  H  H  H  H  H  H  H  H  H  H  H  H  H  H  H  H  H  H  H
```

FIGURE 10–3. Polyethyline chain.

hand is provided by the application of heat and pressure (1,000 atmospheres) or through the action of catalysts. Either way, the double bond is broken (Figure 10–2). The unattached bonds immediately hook up with neighboring molecules. Another and another joins the lineup until chains 100,000 ethylenes long are formed.

Ethylene has become many (poly) ethylenes, polyethylene (Figure 10–3), the most widely used of all plastics. In doing so, it has changed from a gas to a waxy solid—not surprising, considering the similarity to a long-chain member of the paraffin series. The process by which these small molecules, or MONOMERS, join together is known as POLYMERIZATION. The gigantic molecules that monomers create are called POLYMERS.*

Polystyrene

This process is capable with any number of monomers. When benzene and ethylene are chemically combined, they form a colorless flammable liquid with a strong aroma, styrene (Figure 10–4). The molecule resembles both its parents.

Styrene is much less stubborn about polymerizing. When uninhibited (unstabilized against polymerization) and allowed to sit at room temperature for several weeks, styrene will slowly turn into a thick liquid—

---

*Ethylene undergoes ADDITION polymerization, so-called for obvious reasons. Other plastics are made by monomers joining together after they squeeze out such smaller molecules as of water. Nylon is made this way, by CONDENSATION polymerization. In addition, two or more monomers can be polymerized together to form a COPOLYMER.

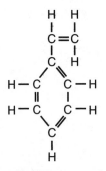

FIGURE 10–4. Styrene (C$_6$H$_5$CHCH$_2$)

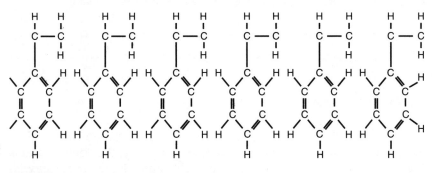

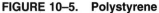

FIGURE 10–5. Polystyrene

a mixture of polymer and monomer—and finally to a glassy solid, polystyrene (Figure 10–5).

If the surrounding temperature is raised to the boiling point of water, styrene will polymerize to polystyrene in a few days. Confined and raised to fire temperatures, the change can take place with violent rapidity. One of the major and continuing causes of explosions during plastics manufacture is uncontrolled polymerization, a reaction that runs away for one reason or another. Some very dangerous chemical compounds are used to control polymerization, to promote and accelerate the process when desired.

Polyethylene and polystyrene polymers are basically threadlike. But other monomers can make linkages in all directions, forming a netlike structure. The difference between a net and a thread can cause many basic differences in the finished plastic. At some time in its life history, as its name implies, a plastic must be soft enough to be worked into a desired shape. A threadlike polymer can be softened repeatedly by heat, hardening into a new shape when cooled. In this, it is not unlike candle wax. It is called a **thermoplastic**. A net is more delicate and set in its ways. Some plastics can be given a permanent shape only once. Reheating will not soften them, for they will char and decompose if too much heat is applied. This type of plastic—for example, a polyester—is known as a **thermoset**. All plastics are one or the other, thermosets or thermoplastics.

MODERN PLASTICS

A modern plastic may not be a single ingredient, but a blending of many substances into a compound.* The properties of a plastic can be altered greatly by the various additives that are thrown into the pot. **Binders** are another name for the resins that have been discussed. They can be synthetic in origin (polyethylene, polystyrene, polyester), natural (amber, shellac), or made from protein or cellulose. **Fillers** can increase the bulk of the plastic at low cost and also may improve the properties. They include such materials as wood flour, cotton fibers, metal wires, clay, carbon blacks, and so on. Plasticizers are added to thermoplastics to develop such properties as flexibility. One plasticizer has already been discussed: camphor. Other additives can include solvents, lubricants, colorants, stabilizers, hardeners, catalysts, and flame retardants. From this witches' brew emerge plastics designed to meet many different and specific requirements. This is one of the reasons why plastics have become so popular. Requirements no longer have to be changed to meet the limitations of existing materials; the reverse can be true.

The Plastics Industry

Today, well over 6,000 companies in the United States make plastics. They fall into three large categories that sometimes overlap: the plastic materials manufacturer who produces the basic resin or compound; the processor who converts the plastic into solid shape; the fabricator and finisher who further fashions and decorates the plastic.

The primary function of the materials company is the formulation of the plastic from basic chemicals. This plastic compound is sold in the form of granules, powder, pellets, flakes, and liquid resins or solutions for processing into finished products. Some plastics materials companies may go a step further and form the resin into sheets, rods, tubes, and film. Some purchase chemicals from which they formulate the plastic resins and compounds. Some make only the compound, purchasing the resin.

From the manufacturer, the resin or compound moves to the processer. Molders produce finished products by forming the molding compound into desired shapes. Extruders may be divided into two groups. The first turns out sheets, film, special shapes, rods, tubing, pipes, and wire coverings. The second includes producers of plastic filaments that can be woven into clothing, rugs, or screening. Film and sheeting processers work by calendering, casting, or extruding the plastic. High-pressure laminators form sheets, rods, and tubes from paper, wood, and cloth that are impregnated with resins. The reinforced plastic manufacturers take liquid resins and

---

*One of the phrases commonly used in the industry is "molding compound," for many plastics are molded into their desired shape.

combine them with such reinforcements as glass fibers, and synthetic fibers to form strong, rigid structural plastic. Finally, the coaters make use of various processes to coat fabric and paper with plastic.

Plastic sheets, rods, tubes, and special shapes are the principal forms with which fabricators and finishers work. Using all types of machine tools, they complete the conversion of these plastics into the finished product. For example, working with rigid sheet material, fabricators form such things as airplane canopies and television lenses. Plastic sheeting and film can be make into such products as shower curtains and rainwear. Many companies are engaged in printing, embossing, metal-plating, or otherwise decorating the plastic.*

Hazards of Plastics Manufacture

A survey of the hazards of the plastics business divides into two parts rather neatly:

1. The multiple problems that surround the creation of a plastic.
2. The properties of the finished product.

First things first, for here lies the greatest likelihood of a fire or explosion. The basic materials—such monomers-to-be as butadiene, styrene, acrylonitrile, phenol, formaldehyde, propylene, ethylene, and the others—may have any combination of reactivity, toxicity, or flammability; some may have more than one or all of these properties. Many of these basic materials must be inhibited to prevent the polymerization that can produce heat and, possibly, explode. Acrylonitrile and phenol are only two of the many monomers that are poisonous. Some plastics, such as the cellulosics, are produced by the action of corrosives. Both flammable liquids and flammable gases are well represented, and with all this present, things can sometimes go wrong. As for example:

> In Louisiana, a $9 million fire occurred in a large petrochemical plant, consisting of a group layout of widely separated block-area units with open air construction, processing basic materials for the production of polyethylene and vinyl chloride resins. Failure of equipment caused a major spillage of ethylene, and the escaping vapor engulfed one of the area units. Within 30 seconds, the ethylene ignited with an explosive roar and quickly involved the process units—the ignition was believed to have originated at cracking furnaces 200 feet away. The deluge (sprinklers) water system was rendered inoperable by the explosion. The fire was finally brought under control by use of hand lines and portable deluge sets, after 3 hours. Practically the entire structure of the process area was collapsed by the intense burning of 250,000 pounds of the

---

*Information in the preceding four paragraphs is taken from "Plastics, the Story of an Industry," a booklet produced by the Society of Plastics Industry, Inc. Those interested in the actual processes of molding, calendering, casting, laminating, reinforcing, extrusion, and fabricating can find an authoritative summary there. Many of the facts on individual plastics are also taken from this source.

product. This operation sustained a shutdown for several months for equipment replacement and structure rebuilding.*

[In another fire] styrene polymer residue, coating the inside supports and roof structure of an empty styrene monomer storage tank of approximately a million gallons capacity, caught on fire as workmen were using welding and cutting torches during a cleaning and altering operation. In approximately 5 minutes (upon the arrival of the fire company) an explosion ripped a part of the cone-shaped roof from one side of the tank. Inaccessibility to the tank interior because of a 30-foot high steel dike limited fire fighting to cooling the tank walls while the fire raging inside burned itself out in about 30 minutes. It was theorized that at welding temperatures, about 2500°F., the styrene polymer accumulations undergo pyrolytic (heat-loosening) decomposition, with some depolymerization and reversion to monomers, which, in turn undergo "cracking" to lighter flammable hydrocarbons. The formation of such hydrocarbons may have been responsible for the explosion.

Monomers are not alone in their combustibility. The entire flaming family is around when a plastic is born: not only gases and liquids, but solids. Solidified resin is often ground or pulverized into finely divided form before use. The fillers previously mentioned include a number of combustible powders, and, as we shall see shortly, many finished plastics are explosive when reduced to dust during processing or fabricating. Flammable liquids are used as solvents in making plastic paints, adhesives, or coatings. Lubricants that reduce stickiness in molding processes are often flammable, especially if the mold is still hot. This list could be extended, but a fire department's inspection of each plastics company in its area is a better way to complete it, for a number of fires have been caused by each of these possible sources.

A $4 million fire in New York was caused when hot resin was discharged into a mixing tank containing 900 gallons of naphtha. The naphtha was vaporized, and it escaped into the building and exploded. Two men were killed.

[In another large fire, this time in New Jersey and costing $8 million] the fire started in the resinous dusts accumulated on the wooden planking over a plastic laminating machine. Employees attempted to extinguish the fire with portable extinguishers. There was a delay in sending in the alarm and the fire spread rapidly between the walls and flooring of the building. The sprinkler system was only partially effective due to lack of water.

[In Massachusetts] a minor explosion in the vicinity of a polystyrene grinding unit was followed by a larger blast in the manner characteristic of a dust explosion. One man was killed and two others were badly burned. It was indicated that the dust housekeeping conditions in the plant were not satisfactory. The plant operated on a round-the-clock schedule.

The initiators or catalysts used to control the polymerization process include some dangerously reactive and highly flammable chemicals, the **organic peroxides**. Among them is a powder that has a burning rate faster than that of cellulose nitrate. This is **uninhibited benzoyl peroxide**.

*Fire histories in this section are taken from AIA, "Fire Hazards of the Plastics Manufacturing and Fabricating Industries," 1963.

The industrial employment of the polymerization process generally takes place in kettles, or autoclaves, where temperatures and pressures may be as high as 1,800°F and 700 psi (982°C and 48 atmospheres). (Molding pressures can run to 30,000 psi [over 2,040 atmospheres].) Murphy's Law should certainly work here: Whatever can go wrong, will. Inadequate controls over any factor may cause the process to "run away": temperature, pressure, reaction times, too much catalyst added too quickly, the improper operation of relief valves, the premature removal of an inert atmosphere. When all this is piled on top of the ordinary fire prevention problems of every industry—improper housekeeping, the presence of unguarded sources of ignition, improper disposal of flammable materials (unless watched, some plants make a habit of using public sewers as a private waste disposal)—one can readily see why fires are not unknown in the plastics industry.

Two more fire histories:

A violent reaction occurred in a 750-gallon reactor during the polymerization of vinyl acetate. Employees tried to control the runaway reaction by cooling but they were not successful. Observation ports of the reactor burst, releasing acetone vapors into the building. Ignition of the vapors caused a violent explosion. One employee was killed. It was reported that the usual pressure relief system did not operate because a 175 psi rupture disc was used instead of a 75 psi disc. Loss: $390,000.

Contents of an electrically heated resin kettle caught fire upon heating the mixture, and subsequently boiled over, spreading the fire to an adjacent warehouse in which a large quantity of resin was stored. Some 250 firefighters spent about 6 hours combating and limiting the fire to the process area, where the fire originated, and to the warehouse. Most of the effort was directed to protecting the exposed tanks and equipment, thus preventing the entire plant from being destroyed. Loss: $700,000.

In addition, **radiation** is used to form cross-linkages, to weave thread-like polymers into nets and improve such properties as the flow rate and tensile strength of a particular plastic. Radiation is also being employed to bind monomers permanently to construction materials.

STRUCTURAL PLASTICS

With billions of pounds of plastics being produced annually in the United States, the hazards of the finished product are of more than passing concern to the fire service. After plastics leave the manufacturer, processor, and fabricator, they are bound to be piled up somewhere in a warehouse. Fortunately the polymerization process has removed much of the toxicity, instability, and flammability of the monomers. Yet, a fire in a warehouse storing several millions of dollars worth of plastic products has often resulted in a total loss.

Plastics are rapidly taking their place alongside wood, glass, stone, and steel as major structural materials. The amount of plastic in a modern building can include that found in: furnishing, cushions, padding, appliances, utensils, rugs, draperies and fabrics, veneer, siding, sealants, insulation, ceiling tiles, floor tiles, wall tiles, cabinetry, decorative panels, door and window trim, lighting diffusers, luminescent panels, window panes, paints and varnishes, structural panels, electrical fixtures, conduits, pipe, wire insulation. . . . In these applications, a dozen or more plastics can be employed. This should cause firefighters to seek the answers to such questions as, How easily will these plastics ignite and sustain combustion? What type of gases and smoke do they develop? How much heat will they generate when burning? How rapidly will a flame spread across their surface? Even if not on fire, how will they react to fire temperatures?

Consider this last question, for it is more important than it appears at first glance. If there is one weakness shared by most plastics, it is inability to withstand continuous temperatures much above 300°F (149°C) and still stay in service. As structural materials, they have a thermal distortion that is of direct concern. What happens when they soften, melt, or deform at fire temperatures?

The plastics industry is well aware of the problem, recognizing its safety responsibility. Maximum service temperatures are invariably included in the specifications of a plastic (see Table 10–3). Only a very few are above 500°F (260°C)—some silicones and cold-molded materials, a fluorocarbon or two. But research is continuing. It has been discovered, for example, that the irradiation of polyethylene will greatly increase its thermal stability. Eventually a solution will be found. Until then, it is almost a certainty that unprotected plastic will generally deform in a fire.

Representatives of the industry sometimes take a positive viewpoint. They point out that light panels, for example, will buckle and fall from their mountings before ignition, thus reducing the possibility of a flame's continuing along the ceiling surface. They also argue that a plastic, used in windows, will distort when heated and provide a vent. Other authorities are less enthusiastic about falling sheets of plastic and sudden venting.

Hazards of Finished Plastics

Most plastics will burn. Everyone agrees on that. But, always excepting cellulose nitrate, the degree of the flammability hazard of plastics is still under active discussion and investigation. Some authorities point to the fact that the self-ignition temperatures of most plastics are hundreds of degrees higher than those of ordinary combustible materials.

Because a solid material generally must be cooled below its ignition temperature before extinguishment, some feel this higher figure means that plastics are both more difficult to ignite and easier to put out. Even though some plastics produce flammable vapors, their flash points are still above the ignition temperatures of wood, paper, and cotton. These authorities also point to tests that show a flame will spread somewhat more slowly on plastic surfaces than on wood, and that plastics will burn

TABLE 10–3. Representative Maximum Service Temperatures for Plastics

	Degrees F	0°	100°	200°	300°	400°	500°	600°
	Degrees C	−18°	38°	93°	149°	204°	260°	316°
Abs plastics								
Acetals								
Casein								
Cellophane								
Cellulose acetate								
Cellulose acetate (foam)								
Cellulose nitrate								
Cellulose triacetate								
Coumarone-indene								
Epoxy (cast)								
Epoxy (foam)								
Ethyl cellulose								
Fluorocarbon (FEP)								
Nylon								

TABLE 10–3. *Continued*

Degrees F Degrees C	0° −18°	100° 38°	200° 93°	300° 149°	400° 204°	500° 260°	600° 316°
Phenolic (foam)							
Polyester (cast)							
Polyethylene (cast)							
Polypropylene							
Polystyrene							
Polystyrene (heat resistant)							
Polystyrene (foam)							
Polyvinyl acetate							
Polyvinyl butyral							
Polyvinyl chloride							
Polyvinyl chloride (foam)							
Polyurethane (foam)							
Urea-formaldehyde							

The temperatures listed here are comparative approximations and can vary widely according to the components in a particular molding compound. In each case, the top of the service range has been taken.

no more intensely than ordinary combustibles. This is especially true if the plastic is treated by a flame retardant or if it is covered by a nonflammable facing. Additionally, incombustible backings can markedly reduce the surface flame spread of thin plastic sheets. The smooth, relatively slick surfaces of a moderately inclined plastic will also tend to shed falling embers and significantly reduce the possibility of ignition.

What tests are being cited? The burning rates of plastics have been investigated many times (see Table 10–4). In 1941, after putting cellulose nitrate in a class by itself, the Underwriters' Laboratories divided plastics into three groups.

Group 1: Those plastics which burn at a rate comparable to cellulose acetate (about like paper) and are more or less completely consumed: some acrylics, polystyrene, and cellulose acetate butyrate.

Group 2: Those plastics that burn with a feeble flame which may or may not propagate away from the source of ignition: urea-formaldehyde, casein, cast phenolformaldehyde and a few others.

TABLE 10–4. The Flash Point and Ignition Temperatures of Certain Plastics and Packing Materials

	Flash point		Ignition temperature	
	Deg. F	Deg. C	Deg. F	Deg. C
Ethyl cellulose	555	291	565	296
Nylon	790	421	795	424
Polyester	750	399	905	485
Polyethylene	645	341	660	349
Polystyrene	680	360	925	496
Polystyrene beads	565	296	915	491
Polyurethane foam	590	310	780	416
Polyvinyl chloride	735	391	850	454
Styrene-acrylonitrile	690	366	850	454
Cotton batting	490	254	490	254
Paper	445	229	445	229
Wood shavings	500	260	500	260

Modern Plastics, July 1961, p. 119.

Group 3: Those plastics which burn only during the application of the test flame: plasticized polyvinyl chloride, a cold-molded plastic and others.

Laboratory testing has continued over the years. The results obtained by the American Society of Testing Materials (ASTM) are often quoted by the plastics industry itself in its literature. A small piece of plastic is clamped into position and set on fire, and measurements are taken on how it behaves. Recently, there has been growing disagreement with the conclusions drawn from these tests. Experts feel that they overly stress low flammability values, and that the phrases used—"self-extinguishing," "slow-burning," and so forth—have little bearing upon conditions experienced in a fire.

The ASTM itself says, speaking on Flammability Tests D-568, D-635 and D-1433: "These methods are intended to provide data for comparing the relative burning rates of plastics in sheet form. Correlation with flammability under use conditions is not necessarily implied."

By now, most fire departments have had enough field experience with plastics in fires to add the weight of our reports on their actual behavior. A tentative agreement has already been established in some areas. Although a plastic may be "slow burning," it can also be extremely difficult to extinguish when glowing combustion burrows deep into a pile of plastic powder, granules, or beads. Some firefighters have reported cases in which "self-extinguishing" plastics burn freely when pre-heated by an approaching fire. There are comments about plastics that melt when ignited, forming a thick liquid that also burns. If such plastics are stored on an open mezzanine, for example, a waterfall of large flaming drops is not only impressive but can add significantly to fire spread or be a sudden shock to any firefighter standing on the ground floor below. The following case histories provide strong evidence of the problems created by burning plastics.

On June 12, 1990 at 9:30 p.m. in Van Buren, New York, a broken six inch natural gas line ignited by an arcing electrical panel started a fire in a 400,000-square-foot manufacturing plant that produces plastic wall decorations, patio furniture, and lawn chairs. The natural gas-fed fire was in close proximity to the 8 inch water line feeding the sprinkler system. The loss of the sprinkler system and water supply greatly taxed the ability of the first arriving units from the Baldwinville Volunteer Fire Department to attack the fire. Within an hour, fire intensity and collapse of the building made defensive operations the only logical choice. It took 30 to 45 minutes before the natural gas line could be shut off. A tunnel existed between the warehouse and the manufacturing portion of the plant. Firefighting crews focused on keeping the fire from spreading through the tunnel into the manufacturing area. Their efforts were successful. In protecting the manufacturing portion of the building, the plant was able to resume normal operations only one day after firefighters left the scene. Being aware of the properties of different materials allows decision makers at the scene of an incident to make quality, informed decisions that result in effective and efficient extinguishment of fire in a safe manner.

In Exeter, Pennsylvania, on October 3, 1989, a fire in a plastics recycling plant involved 20 propane cylinders used to power forklifts in the recycling process. Within 15 feet of the structure on fire was the towns' major lumberyard. The intensity of the fire quickly overran the built-in sprinkler system, failing within ten minutes of the arrival of the first fire units. The roof collapsed early into the incident, rendering the sprinkler system inoperable. Water proved to be ineffective. As water was applied to the burning plastic, it would cool on the surface and encapsulate the burning plastic underneath, thereby not allowing water penetration. Ultimately, foam application proved to be the best tactic.

The burning rate of a plastic can be influenced by several factors. As with other flammable solids, physical form is very important. A thin film will ordinarily burn faster than a thicker piece of the same plastic. For example, tests have shown that a sheet of cellulose acetate will burn at a rate of 4 to 6 inches (10 to 15 cm) per minute, a sheet of cellulose nitrate from 10 to 25 inches (25 to 64 cm) per minute. Compare this with the burning rate of some thinner plastic films, measured in inches per 2 seconds shown in Table 10–5.

There is also the problem of fine division. Plastics are no different from ordinary combustible solids or metals. A plastic that is not especially flammable can become explosive when reduced to dust (Table 10–6) during fabricating or machining or if the resin is produced and handled as a powder.

Foamed Plastics

The surface area and consequent rate of flame spread can be greatly expanded when a plastic is foamed. Resins can be foamed in one of three ways. They can be whipped into a froth just as an egg white can be stiffened. This is called **mechanical** foaming. **Chemical** foaming is accomplished by having gas formed within the polymer by a reaction of some sort—an alkali reacting with an acid, for instance—or by the action of a blowing agent that forms a gas through decomposition. **Physical** foaming comes about through the expansion of compressed gases or volatile liquids incorporated into the resin. Many gases can foam a resin—carbon dioxide, nitrogen, various fluorocarbons, steam, air, and so forth. Ideally, the gas used should have no effect upon the flammable properties of a resin. For this reason, the use of such flammable liquids as pentane, neopentane, and petroleum ether in the formation of polystyrene beads has come under scrutiny. When these beads are exposed to heat, they will expand as the volatile flammable liquids in their structure turn to gas. Naturally, the hazards in fabricating polystyrene beadboard and in the storage of the raw material should be carefully considered during fire inspections and planning, even though the plastic may contain built-in fire retardants.

The use of foamed plastics—rigid or flexible, strong, light in weight, almost impervious to water, with excellent insulating qualities—is one of

TABLE 10–5.   Burning Rates of Plastic Films

Inches/2 Secs.	0	1	2	3	4	5
Centimeters/2 Secs.	0	2.5	5	7.6	10	12.7
Cellophane						
Cellulose acetate						
Cellulose triacetate						
Cellulose acetate butyrate						
Ethyl cellulose						
Nylon						
Polyenthylene, low density						
Polymethyl methacrylate						
Polyvinyl alcohol						
Polyvinyl chloride						

Note: Dotted portion of line represents variation in burning rate among different samples of the same materials. Tests were conducted by clamping thin film of material at a 45° angle (from horizontal) and igniting one edge.

the bright horizons for the industry. Flexible foams are presently being used for cushions and the padding of furniture. Rigid foams are expected to revolutionize concepts in low-cost housing. This deserves some emphasis. Only the flame-retardant grades of foamed plastic should be used as structural materials. It is important that foamed insulations be protected from flames and physical impact damage. Foamed plastics should not be left exposed except in special situations but should be covered with a finish of some kind. When polystyrene foam melts, it shrinks away from ignition sources. Behind noncombustible finishes, it will not support combustion.

Foamed plastics are very susceptible to flame spread. They have a large surface area. Polystyrene and cellulose acetate foams burn at the very respectable rate of 4.5 inches (11.4 cm) per minute. (Other types of foamed plastics include polyurethane, urea-formaldehyde, polyvinyl chloride, epoxies, and the phenolic or silicone resins which are often foamed into place.)

TABLE 10–6. Explosive Properties of Plastic Dusts

	Ignition Sensitivity	Explosion Severity	Ignition Temp. Cloud	Ignition Temp. Layer
Acetal resins	Severe	Strong	824°F 40°C	
Acrylonitrile polymer	Severe	Severe	932°F 500°C	860°F 460°C
Cellulose acetate	Severe	Strong	824°F 440°C	664°F 340°C
Cellulose triacetate	Strong	Strong	806°F 430°C	
Cellulose acetate butyrate	Strong	Strong	770°F 410°C	
Cellulose propionate	Strong	Severe	860°F 460°C	
Coumarone-indene resins	Severe	Severe	1,022°F 550°C	
Epoxy resin, no additives	Severe	Severe	1,004°F 540°C	
Ethyl cellulose, no fillers	Severe	Severe	644°F 340°C	666°F 330°C
Furane resin: phenol-furfural	Severe	Severe	986°F 530°C	
Melamine-formaldehyde	Moderate	Moderate	1,454–1,490°F 790–810°C	
Methyl cellulose, no fillers	Severe	Severe	680°F 360°C	
Nylon polymer	Severe	Strong	932°F 500°C	806°F 430°C
Phenol-formaldehyde	Severe	Strong	1,076°F 580°C	
Polycarbonate resin	Strong	Strong	1,310°F 710°C	
Polyethylene, high-pressure	Severe	Strong	842°F 450°C	716°F 380°C

TABLE 10–6. *Continued*

	Ignition Sensitivity	Explosion Severity	Ignition Temp. Cloud	Ignition Temp. Layer
Polyethylene, low-pressure	Severe	Severe	842°F 450°C	
Polyethylene terephthalate	Strong	Severe	932°F 500°C	
Polypropylene resins	Severe	Strong to severe	788°F 420°C	
Polystyrene beads	Strong	Strong	932°F 500°C	878°F 470°C
Polyurethane resins (foam)	Severe	Strong	950–1,022°F 510 – 550°C	134–824°F 390–440°C
Polyvinyl acetate	Moderate	Weak	1,022°F 550°C	
Polyvinyl butyral	Severe	Moderate	734°F 390°C	
Rayon, viscose	Moderate	Moderate	968°F 520°C	482°F 250°C
Syrene-acrylonitrile copolymer	Strong	Weak	932°F 500°C	

Foam Rubber

While on the subject of foamed plastics, we should not overlook another familiar material which is produced by many of the same methods: foamed latex, also called sponge or foam rubber. Synthetic rubbers, such as neoprene and butadiene-styrene, were developed during World War II to replace the supplies of natural rubber cut off by the Japanese invasion of southeast Asia. During that time, the butadiene-styrene copolymer was known as Government Rubber-Styrene, and it is still known by the initials GR-S and by such trade names as Airfoam or Foamex. Neoprene, created by the DuPont Company from a chlorinated acetylene polymer, now has several trade names. Most foam rubbers are made from GR-S or natural rubber, or from mixtures of the two.

A series of tests at the Underwriters' Laboratories showed that, at temperatures above 210°F (99°C), spontaneous heating of foamed natural rubber took place, accompanied by the production of flammable vapors. If the heat had been kept confined, ignition would have taken place.

Although some types of foam rubber burn vigorously and require sizable amounts of water to extinguish, they burn primarily on the surface. The fire does not burrow into the interior, as it does in a cotton mattress, requiring a complete and tedious overhaul to prevent a rekindle. Quantity storage of most rubbers can lead to a fire of intense heat and rapid spread. But it is the smoke that leaves a lasting impression. Natural rubber is vulcanized by sulfur. Amid the dense black smoke and disagreeable odor of a rubber fire are several toxic gases.

Toxicity of Plastics

When cellulose nitrate burns or decomposes, it releases gases that can be toxic or explosive. How do other plastics behave in this regard? Like everything else, it all depends. Plastics are no different because they are synthetic. Just as cellulose nitrate fumes contain the nitrogen oxides, other burning plastics send forth their own mixture of fire gases. For example, vinyl plastics create significant amounts of hydrogen chloride. The spread of this corrosive gas has been known to cause property damage far beyond the limits of the fire itself. Metallic machinery is neither as valuable nor as vulnerable as your skin and lungs, which must be adequately protected.

Plastics are everywhere. Sometimes the fumes from a vinyl plastic such as polyvinyl chloride (PVC) reach out to injure or hospitalize a firefighter completely unexpectedly, in a routine fence fire where the fence is topped with corrugated PVC. Fortunately, these fumes have an unpleasant, sharp, acrid odor that can't be missed. If concentrations build up to hazardous levels, this odor and a possible skin tingling should not be disregarded.

Many other extremely toxic gases are possible in a plastic fire: ammonia, some of the cyanides, phenol, aldehydes, amines. That they are generally produced in small amounts has led to some conclusions stated in Underwriters' Laboratories Research Report 53, as follows:

> It is recognized that the combustion and thermal decomposition of plastics materials may present a hazard to life under fire conditions, but in the opinion of these authors the chief hazards in this connection are due to the presence of carbon monoxide or atmospheres lacking in oxygen, which is likewise the case where wooden or other cellulose materials are involved in a fire.

Do not, even briefly, consider this as minimizing the dangers of the fire gases that plastics can produce. Rather, it is another recognition of the properties of carbon monoxide, the major cause of fire death in the United States. To a firefighter, anything smoking may be hazardous to health, including plastics.

Some plastics are not far behind rubber in the amount of smoke they will produce. Anyone who has burned a test piece of foamed polystyrene or witnessed a fire in a sizable amount of it will testify to this. The presence of carbon monoxide can even be predicted from the smoke—black, a color created by tiny particles and clumps of unburned carbon, another signal of incomplete combustion. Some plastic fires create a dense, blinding pall that makes any entry a challenge. Many a fire report emphasizes the presence of an astonishing amount of smoke. For example:

> [In an $800,000 fire in Minnesota,] the fire quickly spread through the area. The burning plastics generated huge amounts of black smoke as the fire grew. Firefighters could not find their way into the building to get to the seat of the fire, because of the obscuring, dense smoke.

> [In a polystyrene fire in New York,] the fire spread with such rapidity that hundreds of sprinkler heads were opened and the water supply was insufficient to extinguish the fire. The fire department encountered large quantities of dense black smoke which hampered fire fighting. Flames shot up to a height of 50 feet.

> [At a polystyrene foam fire,] the entire interior of the building was involved when the firefighters arrived. The contents were subject to a heavy smoke damage.

> [At a polyurethane foam fire,] due to the large amount of dense smoke, it was impossible for the fire department to locate or determine the extent of the fire. Sprinklers were, therefore, left on for about an hour and were turned off only after a 20' x 20' hole was cut in the roof to release the smoke.

IDENTIFYING PLASTICS

Even with plastics, the need to identify materials cannot be avoided. Polystyrene does not react like polyvinyl chloride in a fire, and neither resembles cellulose nitrate. Without identification, the potential of a fire is unknown. Instead of developing a fire plan that takes into consideration the cause and possible hazards of the materials involved, firefighters find themselves responding to the effects as they happen. The fire remains a jump ahead, controlling their reactions, instead of the opposite. The positive identification of a particular plastic is difficult. Although approximately twenty groups of plastics will be discussed in the pages immediately following, including some of their uses and trade names, there are reasons why this information will be of limited help.

1. The field grows faster than books can be written; not only are new plastics constantly introduced, but so are improvements upon old varieties.
2. Several different plastics, all looking much alike, can be used for the same purpose. Turning it around, the same plastic can be produced in many forms, from films to shapes to foams, with its properties altered by its structure.
3. Trade names may not be too helpful either. A familiar brand name may reflect only the size of the advertising budget of the parent company, not the amount in use in your particular area. As seen in

the case of cellulose nitrate, many trade names can refer to the same material. It is also possible for the same trade name to refer to completely different types of plastic.

With all these limitations in mind, we will look at what plastics are being manufactured in quantity these days and how we can identify them.

ABS Plastics

Developed in 1948, these thermoplastics combine rigidty with high impact strength. They are made from the copolymerization of three monomers: acrylonitrile (a poisonous, unstable, flammable liquid), butadiene (an asphyxiating, unstable, flammable gas), and styrene (an irritating, unstable, flammable liquid). ABS plastics are available for subsequent processing in the form of powder, granules, or sheets. Trade names: Lustrex, Royalite, or Uscalite. Typical uses: pipe and fittings, wheels, football and safety helmets, refrigerator parts, battery cases, tote boxes, water pump impellers, utensil, tool handles, and radio cases. A moderate fire hazard, ABS plastics are slow burning, do not drip, and have an odor of illuminating gas.

Acetal Resin

Acetal resin is a rigid thermoplastic developed in 1956 for use in fields now dominated by diecast metals. It is extremely tough and has outstanding tensile strength, stiffness, and fatigue life, retaining these properties under adverse conditions of temperature and humidity. It is made from the polymerization of formaldehyde (a toxic, unstable, flammable gas). Acetal resin is produced in powder form, under the trade name of Celcon, a copolymer, and Delrin. Typical uses: automobile instrument clusters, carburetor parts, gears, bearings and bushings, door handles, plumbing fixtures, and moving parts in home appliances and business machines. It burns with a clear blue flame, melting and producing flaming drops.

Acrylics

Acrylics are a group of thermoplastics that combine strength and rigidity with exceptional clarity and good light transmission. In crystal-clear form, acrylic plastics can "conduct" light, pick it up at one edge, and transmit it unseen, even around curves. The various acrylics are made by the polymerization of acrylic acid (a poisonous, unstable, flammable liquid), methacrylic acid (a corrosive, unstable, combustible liquid), acrylonitrile, or other chemically related compounds. Acrylics are available as rigid sheets, rods, tubes, and molding powders for eventual use as airplane canopies and windows, TV and camera viewing lenses, dentures, brush backs and combs, costume jewelry, lamp bases, scale models, outdoor signs and counter displays, automotive tail lights, skylights and paints. The Houston Astrodome is covered by a double layer of acrylic panels. Plastic products can be made several ways, including the forcing of the polymer through the holes of a spinnaret to form a fiber. Trade names for the resin include Lucite and Plexiglas. The fibers have such household names as Acrilan, Dynel, Orlon, and Verel. Acrylics ignite readily, soften, but generally do not drip when burning. The flame is blue with a yellowish-white tip. The smoke may have a sweet or fruity odor.

Amino Plastics

The two major types of these thermosetting plastics are formed by the reaction of formaldehyde with either melamine (a colorless crystalline solid that may release toxic gases when heated) or urea (a white crystal formed industrially by combining ammonia with carbon dioxide, or internally by our urinary systems). Together, these plastics have reached an annual production level of over 400 million pounds (181 million kg). They are color-fast, glossy, scratch-resistant, very hard, and unaffected by many common solvents. Amino plastics are available as foams, white molding powders or granules with such trade names as Cymel or Bettle, or as clear, syrupy resins. Molded products are formed in the presence of acid under the influence of heat and pressure. Typical uses for melamine-formaldehyde: tableware (Melmac), distributor heads, buttons, baked enamel finishes, textile and paper treating. For urea-formaldehyde: lamp reflectors, radio cabinets, plywood adhesives (Griptite and Weldwood). Maximum service temperatures for melamine range between 210°F and 400°F (99°C and 204°C), depending upon the filler. For urea-formaldehyde, 170°F (77°C). The amino plastics will swell and crack when exposed to a flame and produce a smoke that smells like formaldehyde, ammonia, urea, and fish, but they are considered to have a low flammable hazard.

Casein

Many naturally occurring substances such as shellac, amber, pitch, or asphalt are plastics. Plastics can be made from protein, whether the source is animal (hides, bones, hair, horns, or hooves), or vegetable (wheat, peanuts, soybeans, or corn). Casein is the protein found in milk. When it reacts with formaldehyde, it forms a thermosetting plastic (Ameroid) that is available in sheets, rods, and tubes, or as a powder or liquid. Casein was one of the first plastics to be introduced—in 1919. Although strong and rigid, it does not withstand moisture or temperature changes too well. It ignites, swells, and chars when contacted by a flame, producing an odor of burnt milk.

Cellulose Plastics

Cellulose nitrate, formed by the action of nitric acid upon cellulose, was the first acceptable synthetic plastic. Since its creation in 1868, many other chemicals have been put to work upon cellulose. By now, they have formed a large group of usable thermoplastics. Cellulosics are among the toughest of plastics, retaining their lustrous finish under rough usage. But with a few exceptions, they weather poorly. Their maximum service temperatures are rather low, and they must be kept away from alkalis and alcohols. Despite their limitations, the cellulose plastics are available in most of the common forms for a wide variety of uses.

Cellulose Acetate (C/A). Made by treating cotton linters or wood with acetic acid or acetic anhydride, both of which are highly irritating and flammable liquids, C/A is used today for films, such finished articles as washable playing cards and toys, as a textile fiber, or as an expanded foam. Cellulose acetate burns at about the same rate as paper and produces sooty black smoke, a vinegar odor, and a dark yellow flame; it softens and forms bubbles and drops that burn as they fall.

Cellulose Acetate Butyrate (CAB). Made from cellulose after treatment by butyric acid (a poisonous and flammable liquid), acetic acid, and similar anhydrides. One of the more weather-resistant of the cellulosics, it is used in greenhouses and outdoor signs. CAB is considered to be a moderate fire hazard, igniting readily, melting and producing burning drops. The flame is blue with a yellow tip and sparks. Black smoke is created, along with a rancid butter odor.

Cellulose Propionate (CP). Made from cellulose reacting with propionic acid (a highly irritating and flammable liquid) and acetic anhydride, CP is used for the housings of telephones, pens, and pencils and has such trade names as Forticel and Tenite. It burns with a blue flame, a yellowish tip, and sparks; melts, drips fire; and creates a fragrant aroma.

Cellulose Triacetate (CT). Burns at a rate that varies with the plasticizer used in manufacture. It will melt, turn black, and produce an acrid smell. But the flammability rate is quite low, and CT has a relatively high maximum service temperature of 400°F (204°C). It is used in recording tapes, safety glasses, visual aid transparencies, as a textile fiber, and in Safety Base film.

Film made of cellulose acetate burns about like cellophane, a little less than 2.5 inches (6 cm) per second. Cellulose triacetate film burns more slowly, at a rate less than 0.5 inch (1 cm) per second. Compared with the fierce burning of cellulose nitrate film, either acetate burns peacefully. They are far safer in other ways: Ignition temperatures of the acetates are 500 degrees Fahrenheit or 278 degrees Celsius higher than those of nitrate; the decomposition temperature is higher; decomposition fumes are much less toxic, and, very important, decomposition proceeds only when there is a continued source of heat.

Since 1952, all photographic film in this country has been made from Safety Base, cellulose acetate, or cellulose triacetate. However, nitrate film is still around. It is still used abroad, for example; in many movies being shown today in art houses or revivals; and in films dating from before 1952. Converting this nitrate film to Safety Base is not considered worth the cost. Consequently, the problems of nitrate film storage are not over yet.

Recommended safety measures include the segregated storage of nitrate film in separate vaults. Vault doors, reel bands, reel cans, and storage records should be clearly identified with the words "Nitrate Film" in red lettering, "Safety Film" in bright green. When in doubt about a particular reel of film, take a small piece outside and burn it. Doubt will not last long. The words "safety base" are often printed along the edges of the film strip itself to help identify it.

Ethyl Cellulose (E/C). E/C is a molding compound created when a form of cellulose is exposed to an ethylating agent like ethyl chloride (an anesthetic and flammable liquid or gas). It is used in electrical parts, flashlights, and hose nozzles. Igniting readily, it burns with a greenish-yellow flame, has drips that also burn, and creates an odor of burnt sugar. A similar plastic is methyl cellulose, which is also flammable. This plastic can be used to package a powder until the package and the powder dissolve together in water. Trade names: Ethocel or Methocel.

Regenerated Cellulose. Made when a form of wood is treated by sodium hydroxide (a nonflammable, but corrosive liquid or solid), carbon disulfide, and an acid. The familiar brand name for this material is Cellophane, and the uses are equally familiar: bags, pressure-sensitive tape, wrappings, etc. Cellophane burns rapidly, with an odor of burning paper.

Plastic Textiles

When various forms of cellulose are dissolved, processed, and forced through a spinnaret to form a thread, they can be woven into a familiar fabric: **rayon**. Trade names for the various rayons include Arnel (triacetate rayon), Chromspun (acetate rayon), Matesa (cuprammonium rayon), Cordura (viscose rayon), and Fortisan (for rayons spun from cellophane). Plastic fibers are no different from natural fibers. If they are flammable, they will burn at a rate that is partially determined by their weave and length of nap.

As a result of several fires involving synthetic clothing such as "torch sweaters" (which had a history of total consumption by fire within twenty to forty seconds) and shaggy cowboy pants, the clothing industry was forced to develop flame-resistant and flame-retardant clothing. Today, all infant clothing is required to be flame retardant, which explains the high cost of infant wear. In addition, the tendency of firefighters' pants to melt in a fire situation resulted in the development of nomex, kevlar, and PBI. This also resulted in a significant increase in the cost of station uniforms and turnout (protective) clothing.

What is true of clothing is also true of some plastic fibers, such as the acrylics, which are woven into rugs. Rugs with a long fluffy pile have been known to assist greatly the spread of fire in multiple-occupancy dwellings, carrying the flames down hallways, up stairways, and beneath doors to uninvolved areas.

Cold Molded

Cold molded plastics are thermosetting materials produced when an inorganic binder (cement, glass, lime, ceramics) or an organic binder (phenolic resin, bitumin) is mixed with a filler, usually asbestos. The mixture may then be placed in a cold mold under pressure and subsequently baked. Cold molded plastics with organic binders are very difficult to ignite; those with inorganic binders will not burn. Some of these cold molded plastics have very high continuous service temperatures, up to 1,500°F (816°C). Brand names: Tico, Garit, Hemit, or Robit. When heated, they may emit a waxy odor, but there is no sign of melting or dripping.

Coumarone-Indene Resins

One boast of the industry is that these synthetic materials are derived from only six basic sources: air, water, salt, cellulose, petroleum, and coal. Coumarone-indene resins are obtained by heating the light–oil fraction of coal tar. The two liquids are then polymerized with the aid of sulfuric acid, although indene merely needs the help of air and sunlight. Coumarone-indene thermoplastics are used in combination with rubbers, other synthetic and natural resins, or waxes. Trade names include Cumar, Loxite, Neville, and Picco.

Epoxies

First manufactured in 1947, the epoxies are a series of thermosetting resins produced by the copolymerization of epichlorhydrin (a toxic, unstable, flammable liquid) with a phenol or glycol. Epoxies are available commercially as molding compounds, resins, foamed blocks, or liquid solutions bearing such trade names as Cardolite, Durafoam, Fiberlite, and Hysol. With such properties as flexibility, adhesiveness, and chemical resistance, most of the annual production goes into paints, varnishes, and glues. Epoxies ignite readily, charring as they burn.

Fluorocarbons

A cylinder of tetrafluorethylene gas (TFE) unexpectedly polymerized spontaneously into a white powder one evening in 1943. The chemists gathered around. TFE is built about like ethylene, except that fluorine atoms have taken the place of the hydrogen atoms. The gas had formed a solid polymer similar to polyethylene. (Compare Figure 10–6 with Figures 10–1 to 10–3.)

This new plastic, Teflon, had some remarkable properties. It was strong, extremely hard, had a high impact strength, and, for a plastic, was very resistant to temperature extremes. Most fascinating, it was so slippery that practically nothing would stick to it. Teflon could be used to coat frying pans, electric irons, ovens, and other household appliances. Liquids rolled right off this unique plastic that resisted both oil and water. It became possible to create a fluorocarbon with a sticky side and a slippery side. This coating (Zepel, Scotchgard) could be used on fabrics to make them water- and stain-repellent. Several other fluorocarbon polymers and copolymers have emerged over the years. They are used in such industrial applications as valve seats, gaskets, pump diaphragms, linings, tubing, and high voltage insulation.

Fluorocarbon plastics *will not burn*, but they can melt, char, bubble, and decompose when exposed to high amounts of heat. The gases formed during decomposition are very poisonous. Wear self-contained breathing apparatus if fluorocarbon plastics are involved in a fire.

Nylon

Nylon is now the chemical name, the generic name, for a related group of thermoplastic resins also called the polyamides or by such trade names as Tynex and Zytel. It is formed by the condensation polymerization of certain acids with diamines. For example, adepic acid and hexamethyldiamine will condense together after forming a water molecule. This is easier said than done.

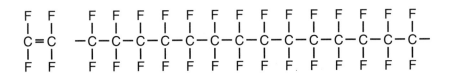

Tetrafluroethylene C_2F_4 Polytetrafluroethylene

FIGURE 10–6. Teflon.

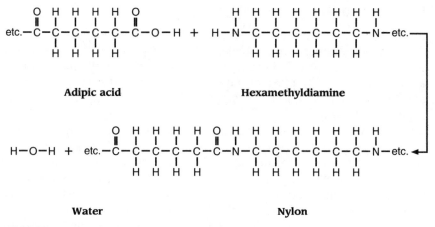

Adipic acid **Hexamethyldiamine**

Water **Nylon**

FIGURE 10–7. Formation of Nylon

Nylon is a very versatile plastic. It is used not only in hosiery and clothing but also in fishing lines, brush bristles, rugs, zippers, washers, gears. Nylon is difficult to ignite. More frequently it will melt, drip, froth, and produce an odor variously described as similar to that of celery, wool, or burning leaves.

Phenolics

There is a little bug, a native of southeast Asia, that at one stage in its life retires behind a resin that it creates from the sap of a tree. Over 150,000 of these lac bugs must be harvested to produce a single pound of shellac. At the beginning of this century, Dr. Leo Baekeland began looking for a synthetic replacement for shellac. He began experimenting with a combination of phenol and formaldehyde. Over 40 years had passed since the introduction of celluloid, and cellulose nitrate was merely the rearrangement of an existing molecule. What Dr. Baekeland eventually created was a brand-new molecule, entirely synthetic, that had no duplicate in nature. He did not stop with the discovery. Dr. Baekeland developed techniques for putting his phenol-formaldehyde resin to work as a phenolic that could be cast—one that could be formed under heat and pressure—and a solution that could be used in laminates. Small wonder the first trade name for this new material honored him: **Bakelite.** Today, additional brand names include Formica, Fiberfil, Resinox. The trade name "Bakelite" now refers to many different groups of plastics.

Phenolics are tough and strong, the real "work horses" of the plastics industry. Typical uses are as distributor heads, telephones, radio and TV cabinets, insulation, home appliances, drawers, dials, handles, knobs, brake linings, bondings, coatings. Several types of phenolics are produced from the reaction of the various phenols and aldehydes. Different fillers, such as mica, will produce compounds which neither burn, melt, nor drip.

Phenolic foams are considered to be self-extinguishing. Thermosetting cast phenolic will ignite, but it burns slowly. It gives off the odor of formaldehyde and phenol. Phenol, also called carbolic acid, is a poisonous flammable solid or liquid.

Polycarbonate Resin

Discovered (also accidentally) in 1957, this transparent thermoplastic (Lexan or Merlon) is the toughest of all plastics. A sledgehammer blow will not shatter it. Thin sheets of Lexan will stop bullets. This polycarbonate has already found such interesting uses as space helmet visors, unbreakable windows, and aircraft gauges. The resin is considered to have a low flammability rating.

Polyesters

Polyesters were first developed in 1942. They are all thermosets with one exception, familiarly known as Mylar recording tape. These resins can be used to impregnate cloth and other materials, to reinforce plastics, or to create synthetic fibers (Dacron and Kodel). They can be cast or molded into automobile bodies, luggage, skylights, translucent roofs, and decorative interior partitions. Cast polyester will ignite, melt at the edges, and produce quantities of black smoke and a faint, sweet odor. Maximum service temperatures range between 250°F and 450°F (121°C and 232°C), depending upon the filler used.

Polyethylene

At last count, a single plastic, polyethylene, made up almost one-third of the industry's total production. The number of uses to which this waxy, lightweight thermoplastic is put would fill many pages. Here are only a few: food and textile packaging, rigid and squeezable bottles, acid-resistant tank linings, bristles, bowls, garbage cans, wastebaskets, toys, electrical insulation, pipe, coatings on paper and other materials, flexible ice cube trays, glasses, dishes, carboys, rain capes, meteorological balloons, greenhouses, silo covers, cable jacketing, moisture barriers under concrete, walls, cordage, filter cloths, upholstery, and artificial flowers and grass. Trade names include Polyfilm, Spunglow, Visqueen, and Surlyn (a new see-through type). One reason for the popularity of polyethylene is the way it can be altered to fit a particular need. The polyethylene chain can be made to branch or cross-link. When this happens, the plastic will change its density, flexibility, and resistance to heat. But even at its best, it should be kept out of ovens. The maximum service temperatures range between 165°F and 300°F (74°C and 149°C). Polyethylene melts as it burns, and the drippings continue to burn. Once ignited, it burns rapidly with a blue flame tipped with yellow, giving off an odor like that of burning wax.

Polypropylene

First introduced into the United States in 1957, polypropylene is formed, as might be expected, by the polymerization of propylene (an anesthetic, flammable gas). It has better heat resistance than polyethylene and can be used in sterilizable bottles and containers, for molded products (Moplen) that have an unusual chemical resistance, in fibers (Firestone), and in transparent films (Propylene) that are impermeable to vapors and gases. This thermoplastic ignites, burns slowly, melts and drips, and produces an odor not unlike that of heated asphalt.

Polystyrene

I. G. Farben* chemists first made polystyrene in 1929. It can be processed into rigid forms (Styron, Pliolite, Cerex), into films (Tricite, Styrofilm), or into foams (Styrofoam, Scotbord). Molded polystyrene is widely used in containers, battery cases, radios and the like; the foam in toys, insulation, counter displays, and package cushioning; or foamed-in-place. The maximum service temperatures for polystyrene are rather low. Even the so-called heat-and-chemical-resistant types do not withstand continuous exposure to temperatures much above 200°F (93°C). Polystyrene ignites easily, softens, bubbles, and burns with the formation of impressive amounts of black smoke. Burning polystyrene has an odor variously described as that of illuminating gas or marigolds. Polystyrene is a thermoplastic.

Polyurethane

During World War II, the Luftwaffe needed strong wing tips, and with the way the Spitfires were acting, they needed them at once. German technology provided a short-term assist by developing polyurethane foam. In recent years, industrial interest in these plastics has expanded rapidly, as the number of trade names testifies: Curifoam, Eccofoam, Genfoam, Pliofoam, Polyfoam, Polylite, Polyrubber, Scotchfoam, Selectrofoam, Stafoam, Vibrafoam, and many others. Flexible urethane foams have been used for some time in cushions, mattresses, rug underlays, sponges, mats, and so on. Rigid urethane foams are now coming into their own as highly efficient insulators. They are the reason why the walls of our refrigerators are thinner than they were formerly. In addition, they can add strength to structures without a great deal of weight, are highly buoyant, and can be foamed in place. Urethane foam reinforced by a polyester (Corfam) can even substitute for leather. Or, take a good look at a cigarette filter.

---

* I. G. Farben is not a person's name. It stands for the German "Industrie Gesellschaft Farben," Industrial Dye Corporation. This was a pre–World War II industrial combine or cartel that produced everything from photographic film (AGFA) to medicinal products.

One of the leading manufacturers of the raw materials, Union Carbide Company, has several interesting comments about rigid polyurethane foam.

Rigid urethane foam is formed by the reaction of two liquids in the presence of a gas-producing blowing agent, usually carbon dioxide (or a fluorocarbon). As the chemical reaction occurs, heat is generated and the blowing agent vaporizes to form tiny bubbles in the thickening plastic. In less than 2 minutes, the foam expands to its full height, and sets. What is essentially (formed) is one giant cross-linked molecule containing entrapped bubbles of gas.

Rigid urethane foam has inherent adhesion properties as it foams. This adhesion is so tenacious that rigid urethane foam can be applied by spraying. A foam-producing operation should be well ventilated. Full employee protective equipment is needed and it is frequently wise to use chemical cartridge respirators. Cured rigid urethane foams are inert and do not represent a toxicological hazard.

Polyruethane foams are combustible; however, rigid urethane foams can be made so that they will not support combustion, or will not burn through when heated with a propane torch. Such resistance to burning is achieved at either increased cost or sacrifice in other properties. It is always desirable to test rigid urethane foam under the actual condition to be encountered in use. Like many plastics, urethane foam has a softening point. This softening point appears to be near 250°F. (121°C.).

Polyvinyls

The thermoplastic group of plastics known as the polyvinyls or vinyls are another member of the billion-pound-a-year club, ranking second in amount of use only to polyethylene. They are produced in several forms: flexible, rigid, as fibers, or as foams. All types of vinyls are tough and strong. Although they will stand up to household temperatures, they should be kept away from any contact with heat sources. Most of them are slow burning; a few will not sustain combustion unless in contact with a flame. Table 10–7 lists some of the uses, trade names, and flammability characteristics of the common vinyl plastics.

Silicones

Developed in 1943, the silicones—so called because their structure is based upon silicon rather than on carbon—are available as resins, coatings, greases, fluids, and silicone rubbers. Their degree of flammability is determined by their filler—either slow burning or completely nonburning. Up to now, the big outlets for the silicones have been adhesives, cosmetic ingredients, and such electrical parts as switches, insulation, coils, and power cables. Typically, research has uncovered some exciting new uses for them. Silicones are significantly more resistant to heat than other plastics. New types have been produced which withstand the tremendous frictional heat of rocket re-entries by slowly decomposing, layer by layer. The process is called ablation. Silicone rubbers in extremely thin sheets will allow the passage of oxygen, but not water. Some day, undersea

TABLE 10–7. Vinyl Plastics

	Maximum Service Temp.	Typical Uses and Trade Names	Fire Characteristics
Polyvinyl acetal	125–150°F 52–66°C	Adhesives, inks, plastic wood. Rigid: INSULAR, LEMAC	Low flammability: ignites, softens, yellow flame, sooty smoke, vinegar odor. The vinyl acetate monomer is an unstable flammable liquid, with low toxicity.
Polyvinyl alcohol		Water-soluble packages for dyes, soaps, detergents. Rigid: ELVANOL, RESISTOFLEX	Moderate flammability: ignites readily, softens, melts, spatters, blisters. Sweet floral odor.
Polyvinyl butyral	100–140°F 38–60°C	Inner layer in safety glass. Rigid: BUTACITE, SAFLEX	Low flammability: ignites, softens, melts, drips; blue flame with a yellow tip; rancid butter odor. The vinyl butyral monomer is a fairly toxic flammable liquid.
Polyvinyl chloride (PVC)	120–175°F 49–79°C	Automobile seat covers, flooring, upholstery, wall and floor coverings, corrugated fence toppings, tarpaulins, packaging. Rigid: VELON, RESINITE Films: NAUGAHYDE, PRESTOFLEX, KOROSEAL Foams: U.S., VINACEL, VINYLFOAM	Low flammability: burns in contact with a flame. Softens under heat and produces a toxic white smoke and a chlorine odor. When encountered (see list of uses) respiratory protection must be worn. The vinyl chloride monomer is an anesthetic, unstable gas which is frequently found liquefied.
Polyvinyl chloride-Polyvinyl acetate copolymer	150–175°F 66–79°C	Textile fibers, baby pants, shoes, rainwear, shower curtains, phonograph records, coatings. Fiber: VINYON Films: VINYLITE, FABTEX, BAKELITE	Low flammability: difficult to ignite, softens, chars, bubbles; smoky yellow flame; noticeable odor of acrid hydrochloric. Respiratory protection must be worn.
Polyvinylidine chloride	160–200°F 71–93°C	Upholstery fabric, screening, draperies, food wrap. Saran Wrap is a copolymer with PVC. Rigid: GEON Fiber: SARAN, VELON	Low flammability: difficult to ignite, softens, chars, odor of chlorine. Respiratory protection must be worn. The vinylidine chloride monomer is a moderately toxic, unstable, flammable liquid.

cities may be enclosed in these plastic films, or perhaps they can be used to replace lung tissue. Like the present, the future of humankind, wherever we go, may well be packaged in plastics.

SUMMARY

The brave new worlds of plastic are not the real concern. What confronts us is sorting out and identifying individuals and groups amid the avalanche of polymers so that some sort of an intelligent fireground decision can be made. Just as a golf course is the worst place to learn how to play the game, a fire scene is not the time to learn suddenly that a plastic produces toxic smoke or that its burning rate is beyond expectation, or even reason. Firefighters must find out what *may* happen before it does! This is why the major types of plastics, how and why they are used, what trade names they bear, and how they will react in a fire have been presented in this chapter. Planning begins with identification.

Although much is still unknown about the behavior of plastics in a fire situation, we do know that many of them are flammable. Their rate of burning can be changed by their form, and fireground tactics must take this into account. Toxic hazards exist: carbon monoxide primarily, but also other unfriendly gases and the possibility of dense concentrations of smoke. These are considerations that demand a place in fireground size-up. Additionally, certain resins, such as the epoxies and the isocyanates (polyurethane), become more toxic when heated. A good rule of thumb is to suspect all plastics of plotting an assault upon the respiratory system until they are proven innocent. Firefighters should protect themselves at all times with at least a self-contained breathing apparatus. (Incidentally, if a corrosive gas is released by a plastic such as one of the chlorinated vinyls, warn the owners that metallic equipment may have to be promptly cleaned or neutralized. Their insurance company will be grateful.)

With the exception of cellulose nitrate, shipping and storage regulations reflect the conclusion that plastics have a degree of hazard similar to that of ordinary Class "A" materials. Why, occasionally, do they cause such trouble? All too often, the answer is mishandling. The industry itself has stressed this fact for years. A plastic is designed with a set of properties that must be respected. Time and again plastics are misused in ways that their characteristics make dangerous. Consider the following examples:

1. A warehouse owner allows great heaps of polystyrene beads to lie about in open bins separated from an ignition source only by good fortune.
2. A builder doesn't use fire-retardant grades of foamed insulation or leaves them exposed to physical damage and flame. Runs of flammable plastics are not interrupted by the separation needed to reduce an area to hand-line size.

3. A woman puts out a glassy-looking thermoplastic coaster, and a guest mistakes it for an ashtray. Her husband, the home handyman, installs thermoplastic wall tiles around his stove. He is so proud of their decorator colors that he never notices them beginning to curl toward the oven.

And don't think that firefighters are exempt. They know that water is recommended for a plastic fire. But it must be used wisely. They are not above directing a straight stream into a pile of plastic dust and powder, causing the formation of an explosive cloud. Also, a plastic fire is very likely to burrow into a pile, awaiting a time to rekindle. The premature closing of sprinkler systems and hose lines has been a source of regret for many fire officers.

REVIEW QUESTIONS

1. What are two names often given to cellulose nitrate plastics?
2. Discuss the rate of combustion of cellulose nitrate.
3. What are some visible signs that decomposition of cellulose nitrate is taking place?
4. Name three gases produced by decomposing cellulose nitrate. How many are flammable?
5. What should be done if cellulose nitrate is spilled inside a truck in transit?
6. Name three ways in which cellulose nitrate can cause an explosion.
7. What is the recommended extinguishing agent for cellulose nitrate?
8. What is a monomer? A polymer?
9. What is the difference between a thermoplastic and a thermoset?
10. Why is the maximum service temperature of a plastic important to a firefighter?
11. Name a plastic that can melt in a fire and produce flaming drops.
12. Name a plastic that produces great quantities of black smoke when on fire.
13. Name a plastic that can produce a corrosive gas when heated or on fire.
14. What identification measures are recommended for cellulose nitrate film?
15. Discuss some of the hazards of the spontaneous heating of foam rubber.

BIBLIOGRAPHY

1. "Giant Molecules," Time-Life, Inc. (1967). An extremely interesting general survey of the plastics industry.
2. Basic source for information about plastics: *Modern Plastics Encyclopedia* (1967).
3. "'Plastic Fires' Create New Hazards for Both Firemen and Public," *Journal of the American Medical Association* (December 22, 1975).
4. "Sensory Irritation Evoked by Plastic Decomposition Products," *American Industrial Hygiene Association Journal* (October 1974).

5. "Safety and Plastic Conduit," *New England Electrical News* (May 1974).
6. Union Carbide Company: "Rigid Urethane Foam" (1965); "Bakelite Phenoxy Resins" (1965).
7. "Plastics as Materials of Construction," Dow Chemical Company, an address given by R. W. Theobald (1966).
8. "Plastics," *Handbook of Industrial Loss Prevention*, 2nd Edition, Factory Mutual Division, Chapter 63, (New York: McGraw-Hill Book Company, 1967).
9. National Fire Protection Association, 1 Batterymarch Park, Quincy, MA.
   Standard No. 40, "Storage and Handling of Cellulose Nitrate Motion Picture Film."
   Standard No. 40E, "Storage, Handling, and Use of Pyroxylin Plastic in Factories."
   Standard No. 40E, "Storage and Sale of Pyroxylin Plastic in Warehouses, Wholesale, Jobbing, and Retail Stores."
   Standard No. 231, "Recommended Safe Practices for General Storage."
   Standard No. 654, "Prevention of Dust Explosions in the Plastics Industry."
   Handbook: Section 6, Chapter V; Section 7, Chapter VI; Section 8, Chapter VIII.

FURTHER READINGS

1. Edwards, R., Edwards, D.; *Fire Chem II*, S.A.F.E. Films, Inc., 3326 Bentwood S.E., Grand Rapids, MI., 1989, pp. 194–225. Text supported with videotapes.
2. Wallace, Deborah, "In the Mouth of the Dragon: Toxic Fires in the Age of Plastics," Garden City Park, NY: Avery Publishing Group, Inc., 1990.
3. Fire, Frank L., "Combustibility of Plastics," Fire Engineering Books & Videos, P.O. Box 21288, Tulsa, OK, 74121-9971.

VISUAL AIDS

Plastics: The Burning Issue—Vinyl: Trial by Fire, Hazardous Materials Library, FEMA Region IX, Presidio of San Francisco, FL5-4-52A. (60 minutes, VHS format)

DEMONSTRATION

With the cooperation of a local industry, you can present to your class two very interesting demonstrations. The burning rate of cellulose nitrate is an eye-opener. Unless you use very small amounts, this demonstration may have to take place outside. Its combustion rate and the way a fire will spread are clearly demonstrated if you make a short trail of the cottony material.

Polyurethane foaming resins are now sold in do-it-yourself packages. It is a simple matter to combine the two reactive liquids in front of the class. Do it inside a large Pyrex glass container, for the increase in volume is surprising.

Oxidizing Agents

11

After considerable discussion of flammable fuels of every description and origin, it is now time to change the subject. From time to time, the subject of oxygen has come up: as part of the atmosphere or the fire tetrahedron, as a pressurized or liquefied gas. The existence of a number of **oxidizing agents** was commented on in passing. As their name implies, these are substances containing oxygen atoms that can be released. If this loosening of insecure chemical bonds takes place during a fire, some interesting consequences can ensue. Now is the time for a more thorough look at these oxidizers.

A problem arises immediately. The hazards of oxidizing agents vary. Which of them should be included here, which should be reserved for another, more appropriate, time and place? For example, although ammonium perchlorate and hydrogen peroxide are part of this chapter, they could just as logically have been included under "explosives" or unstables." The choice depends on the particular properties to be considered. Often, there is a thin line between an oxidizing agent and an explosive. As shall be seen, every one of these oxidizers is capable of crossing this line in one way or another, of exploding under the right (wrong) circumstances.

This chapter deals with oxidizing agents, but not those that are also flammable. One troublemaker like this has been discussed previously, cellulose nitrate, and there are others, such as the organic peroxides. (These are known to the fire service as S.O.B.'s, which undoubtedly stands for self-oxidizing burnables.) When the temperature increase caused by decomposition is added to such substances, all sides of the fire tetrahedron are present in one unstable package.

Note especially that none of the oxidizing agents in this chapter is considered combustible. Just as a fuel must be oxidized before combustion takes place, these oxidizing agents must be fueled. Rather arbitrarily, three loosely related groups of oxidizers have been selected for investigation. Their names sound like a zoo of strange chemical animals, awaiting an opportunity to bite.

1. The chlorites, chlorates, and perchlorates.
2. Some "pers": the permanganates, persulfates, and peroxides.
3. The nitrates and nitrites.

HOW TO BUILD AN OXIDIZER

When one of two words in the name of a substance changes, two completely different chemical compounds or groups of compounds are the result. The difference between "cellulose *nitrate*" and "cellulose *acetate*," or even between "**organic** peroxide" and "**inorganic** peroxide," cannot be ignored. However, look closely when identification hinges upon the change of a single interior letter. Notice the close resemblance between these names of oxidizing agents: sodium nit*rate* and sodium nit*rite*; or even worse, sodium chlo*ride*, sodium chlo*rite*, and sodium chlo*rate*. Does it really matter? Very much so. For, in those sodium compounds, a change of one letter marks the difference between the salt sprinkled on scrambled eggs and an oxidizing agent that can scramble you. All this leads us back to chemistry.

Groups of atoms inside a molecule tend to operate as a unit. This section is all about these "units," for it is within these clusters of atoms that available oxygen is waiting to make an ignition easier, a fire more intense, and/or an explosion possible.*

Cellulose nitrate provided an introduction to the nitrate group: NO_3. A nitrogen atom can also combine with two oxygen atoms to form a nitrite: NO_2. Similarly, chlorites (ClO_2), chlorates (ClO_3), and perchlorates (ClO_4) have available oxygen. In the case of oxidizing agents, these groups of atoms are generally, but not always, attached to some metal: barium, calcium, lead, magnesium, potassium, strontium, silver, sodium, and others. Putting them together with sodium, for example, gives the following results:

Sodium chloride	NaCl
Sodium chlorite	$NaClO_2$
Sodium chlorate	$NaClO_3$
Sodium perchlorate	$NaClO_4$
Sodium nitrite	$NaNO_2$
Sodium nitrate	$NaNO_3$**

*These groups have several names. Sometimes they are called radicals; sometimes, because of their net electrical charge, they are known as ions, cations, or anions.
**The chemical suffixes *-ite* and *-ate* refer to groups that contain oxygen. An *-ate* has one more oxygen than an *-ite*. Incidentally, the word *chloride* has nothing to do with oxygen; it merely informs you that chlorine is present.

Common sense dictates that the more oxygen a group contains, the more it can release to a fire, and the greater its hazard should be. This is true only some of the time. A nitrate is more hazardous than the equivalent nitrite. Sodium chlorite is certainly more dangerous than sodium chloride, which is not an oxidizer at all but common table salt. Yet sodium perchlorate, with four oxygen atoms, is more stable than sodium chlorate, which has only three. And most authorities consider sodium chlorite every bit as dangerous as sodium chlorate.

Chlorites, Chlorates, and Perchlorates

As always, our first concern is with identification. Although there are some variations in uses and properties (refer to Table 11–1), most of the chlorites, chlorates,and perchlorates look about the same. It is difficult to tell one of these water-soluble white powders or crystals from another. They all bear the same yellow DOT label with black lettering and symbol: oxidizer; or organic peroxide, if the product is organic.

To repeat and emphasize a point, these salts will not burn. They give off oxygen, especially when heated, and thereby increase the combustion rate of any flammable material nearby. This phrase "increase the combustion rate" is too easily passed over unless one has witnessed a demonstration of what it means. On a small scale, a cloth soaked in a chlorate solution, allowed to dry, and set afire is here one second, gone the next, in a burst of flame. On a large scale, a fire involving these agents becomes extremely dangerous. The speed of burning, the amount of heat produced, and the rate of fire spread are above all normal expectations, and explosions are possible. Many commercial explosives contain an oxidizer. Some explosives, with a few refinements, are nothing more than an intimate mixture of a finely divided fuel and an oxidizing agent. Chlorates are used in priming caps. Large percentages of potassium chlorate and ammonium perchlorate are in an explosive called Cheddite. All by itself, ammonium perchlorate has already gone a long way down the road toward being an explosive. It requires only a little encouragement to finish its journey.

Perchloric acid, $HClO_4$, has a molecule completely filled with oxygen. Many other oxidizers can also be identified by this prefix: perchlorates, perborates, persulfates, permanganates, and others. Representatives of these groups can be found in Tables 11–1 and 11–2.

Chlorites, chlorates, and perchlorates are potentially explosive in contact with many combustibles, including sulfur, organic materials, and metal powders. These mixtures not only ignite very easily and burn intensely but can explode if subjected to shock, friction, or a heat increase. Moreover, these oxidizing agents decompose at different temperatures. The decomposition may occur with explosive rapidity, or containers may burst simply because of the vapor pressure buildup caused by decomposition gases. Explosions also happen because of chemical reactivity with nonflammable

TABLE 11-1. Properties and Uses of Chlorites, Chlorates, and Perchlorates

	Properties	Uses
Ammonium perchlorate NH_4ClO_4	Powerful oxidizer. Very sensitive to explosion when shocked, exposed to heat, or contaminated. Decomposes upon heating.	Explosives, fireworks, etching, and engraving.
Barium chlorate $Ba(ClO_3)_2$	*Poisonous.* Gives off oxygen at 482°F (250°C). Melts at 777°F (414°C).	Pyrotechnics (green fire), dye fixing, explosives, manufacture of other chlorates.
Barium perchlorate $Ba(ClO_4)_2 \cdot 3H_2O$	*Poisonous.* Melts at 941°F (505°C).	Drying agent for gases.
Calcium chlorate $Ca(ClO_3)_2 \cdot 2H_2O$	When rapidly heated, melts at 212°F (100°C). Hygroscopic.	Photography, pyrotechnics, dusting powder to kill poison ivy.
Magnesium perchlorate $Mg(ClO_4)_2$	Decomposes when heated above 482°F (250°C). Dissolves in water and creates considerable heat. Skin and eye irritant.	Drying agent for gases.
Potassium chlorate $KClO_3$	*Poisonous.* Decomposes at 752°F (400°C), giving off oxygen. Melts at 695° F (368°C). Explodes with sulfuric acid contact.	Oxidizing agent, explosives, matches, fireworks, percussion caps, medicines, dyes, bleach.
Potassium perchlorate $KClO_4$	Decomposes at 752°F (400°C). Irritating to eyes and skin.	Explosives, medicine, oxidizer, photography, pyrotechnics, fuses.
Sodium chlorate $NaClO_3$	*Poisonous.* Liberates oxygen at 572°F (300°C). Melts at 491°F (255°C).	Oxidizer, matches, explosives, leather tanning, weed killer, bleach.
Sodium chlorite $NaClO_2$	*Poisonous.* Decomposes with heat at 347°F (175°C). Forms explosive and poisonous gas in contact with mineral acids.	Improving taste of water, bleach for textiles, paper, oils, waxes, shellac, varnishes, straws.
Sodium perchlorate $NaClO_4$	Decomposes at 266°F (130°C). Melts at 900°F (482°C). Irritating to eyes and skin.	Explosives, jet fuels.
Strontium chlorate $Sr(ClO_3)_2$	Decomposes at 248°F (120°C).	Pyrotechnics (red fire).

TABLE 11–2. Permanganates, Persulfates, Perborates (Oxidizers)

Oxidizer	Properties	Shipping And Uses
Ammonium persulfate $NH_4S_2O_8$	White powder. Irritating to skin and mucous membranes. Yields toxic sulfur oxides when heated. Decomposes.	Shipped in containers from bottles to 300-pound (136-kg.) barrels. Uses: oxidizing agent, bleach, dyes. DOT yellow label.
Potassium permanganate $KMnO_4$ purple salt	Dark purple crystals with a blue sheen. Decomposes at 464°F (240°C). Explodes in contact with organic materials, and with hydrogen peroxide either in solution or in dry state. Dangerous fire hazard. Ingestion of salt can cause death.	Shipped in containers from bottles to 500-pound (237-kg.) drums. Uses: wood preservative, deodorant, bleach, disinfectant (athlete's foot), dyes, absorbent for poison gas. DOT yellow label.
Potassium persulfate $K_2S_2O_8$ Anthion	White crystals. Decomposes and frees oxygen at 212°F (100°C) when dry. In solution, frees oxygen at room temperatures. Skin irritant. Yields toxic sulfur oxides when heated.	Shipped in containers from bottles to 300-pound (136-kg.) drums. Uses: bleach, oxidizing agent, photography, soap manufacture, antiseptic. DOT yellow label.
Sodium perborate $NaBO_2$	White crystals. Stable in cool, dry air, but will decompose and give off oxygen when heated or moist. Nonhazardous unless mixed with highly combustible or reactive compounds.	Shipped in cartons, kegs, barrels. Uses: dyes, bleaches, detergents, germicide, deodorant, neutralizing cold wave preparations. No shipping regulations.
Sodium permanganate $NaMnO_4$	Reddish-black crystals or powder. Decomposes. Dangerous fire hazard.	Shipped in containers from bottles to 300-pound (136-kg.) drums. Uses: oxidizing agent, disinfectant, antidote for certain poisons. DOT yellow label.

materials, such as certain acids and ammonium salts. Broadly speaking, chlorates are considered a bit more unstable than perchlorates.

One fire involving sodium chlorate occurred in a chemical laboratory of a university. Five employees were removing approximately 450 pounds (204 kg) of a chlorate from the basement of a building where it had been stored for years. Authorities had stressed the need for careful handling. The chlorate, in a number of metal and wooden boxes, was being loaded onto a platform truck when there was a sizzling sound, then flame and smoke from one of the cartons. Two explosions and a flash fire followed. Four persons were killed; three more, injured. A number of students, teachers, and firefighters suffered from smoke inhalation.

From past discussion, it is easy to see why flammable materials should *never* be stored in the same area with oxidizing agents. Storage buildings should be of fire-resistant or noncombustible construction. Often overlooked is the obvious fact that wooden floors will burn, and so will wooden pallets and surrounding refuse. If oxidizing agents are spilled, they must be cleaned up promptly—carefully avoiding the use of a combustible sweeping compound—and disposed of carefully. Containers, whether glass bottles or metal drums, should be protected in storage.

The amount of chlorite, chlorate, or perchlorate involved in a fire dictates fire strategy. Sizable quantities require the use of large streams from a distance with firefighters in explosion-protected positions. Be careful not to knock containers about with hose streams. Smothering agents such as carbon dioxide and steam are useless against a fire that has a nonatmospheric source of oxygen. If ammonium perchlorate is being threatened by a large fire, it may be necessary to evacuate the surrounding area. Because the gases of decomposition are toxic, firefighters should be protected with self-contained breathing apparatus. Several of these salts are also poisonous. (See Table 11–1.)

What's a "Per"?

When the name of a chemical compound leads off with the prefix *per-*, it means that its molecule has a distinctive structure. The molecule can be so saturated with an element that there is no more room for additions. For instance, perchloroethylene, the common dry-cleaning fluid, and perchloromethane (carbon tetrachloride) are loaded with chlorine (See Figure 11–1).

Peroxides

Two oxygen atoms can hook together in the manner shown in Figure 11–2. Such a lineup is called a "peroxy group," and the compound that contains it a "peroxide." When these oxygens are surrounded by groups of atoms containing carbon, the resulting compound is a highly flammable and unstable organic peroxide: benzoyl peroxide, for example, or methyl ethyl ketone peroxide. If the peroxy group is joined to a metal such as sodium or barium, or if two hydrogen atoms are linked to the pair of oxygens, an inorganic peroxide is created.

HYDROGEN PEROXIDE

The next time someone you know decides to bleach his or her hair, don't mutter about vanity. Take a good look instead at the bottle of peroxide, for it is a remarkable liquid. Common household peroxide is a 3 percent

Perchlroethylene   **Perchloromethane**
  C_2Cl_4 CCl_4

FIGURE 11–1.   "Per" compounds.

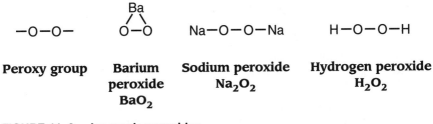

Peroxy group   **Barium**   **Sodium peroxide**   **Hydrogen peroxide**
                    peroxide Na_2O_2 H_2O_2
                    BaO_2

FIGURE 11–2.   Inorganic peroxides.

solution. This means that 3 percent, by weight, of H_2O_2 has been dissolved in a large percentage of water. Hydrogen peroxide is completely soluble in water. At low concentrations, it forms a colorless, odorless liquid that is difficult to tell from pure water, except that it begins to show some of the characteristics that make it hazardous. As its users are well aware, hydrogen peroxide can be used for a bleach. What they may not know is that the bleaching is caused by a controlled oxidation process that burns hair to a lighter color. Hydrogen peroxide is a powerful oxidizer, but it is slow acting and therefore less damaging when it is used to bleach hair or such fabrics as wool, silk, linen, and fur.

How does it work? Hydrogen peroxide is nothing more than water with an extra oxygen atom jammed in.* If given the slightest chance, it will revert to water by decomposing.

$$2\ H_2O_2 \longrightarrow 2H_2O + O_2 + Heat$$

This reaction has two interesting consequences. First, it is **exothermic** (produces heat); second, it frees oxygen. But an overwhelming percentage of a 3 percent solution is water. The heat is easily absorbed and not too significant. Neither is the amount of oxygen set loose. Decomposition

---

*Technically, water can be called hydrogen oxide. The usefulness of this bit of information is limited, except to warn you that some pranksters have been known to include the location of large hydrogen oxide tanks on their industrial inspection reports, giggling as the fire inspectors dig into their chemical dictionaries looking for this dangerous-sounding compound.

generally proceeds slowly, less than 1 percent a year under normal conditions, unless you leave the cap off the bottle. It would take a long time for household peroxide to turn back into pure water.

But if you hold a bottle of peroxide, the heat of your hand will soon release visible oxygen bubbles from the liquid. Your hand is about 20 degrees hotter than normal room temperatures. Hydrogen peroxide will decompose 1.5 times faster for each temperature rise of ten degrees Fahrenheit (5 1/2 degrees Celsius). Also, there is a reason why peroxide bottles are invariably made of colored glass, dark blue or brown; decomposition of the liquid also increases when it is exposed to light.

Hydrogen peroxide is often used as a mild antiseptic, although it is relatively poor for the purpose. When put on a cut, it will foam. This foam is made of tiny oxygen bubbles. Foreign material in the blood causes an even more rapid breakdown of the peroxide than would be caused by exposure to body temperatures alone. This "blood catalase," as it is called, is not unique in its ability to increase the decomposition rate of hydrogen peroxide.

What all of this means is that the rate of decomposition is variable and subject to change if H_2O_2 is contaminated, the temperature is increased, or even if light is allowed to strike the liquid. Despite this, hydrogen peroxide would still be considered a fairly innocuous liquid, an acceptable plaything, were it not for the fact that its reactions have been described in the lowest of concentrations.

During the discussion on ammonia it was learned that industrial concentrations of ammonium hydroxide were far stronger and more dangerous than household ammonia. This is also true of hydrogen peroxide. To learn how a higher concentration of hydrogen peroxide accentuates its hazards, look at some of its common industrial strengths.

Toxicity

There is a warning inside most packages of hair lighteners: "This product contains ingredients which may cause skin irritation on certain individuals, and a preliminary test should first be made (on a patch of skin). This product must not be used on eyelashes or eyebrows. To do so may cause blindness."

When the concentration of H_2O_2 in water reaches 8 percent, a DOT yellow label: oxidizer, is required during shipment. In still higher concentrations, toxic hazards continue to increase. Vapors and mists become more and more irritating to the eyes, nose, and throat. The liquid can cause severe skin burns and blisters, especially in concentrations above 35 percent.

However, an immediate antidote for this corrosiveness is inside a fire hose. Remember, hydrogen peroxide is *completely soluble* in water. Dangerous concentrations can be reduced to nontoxic levels when enough water is added. Firefighters splashed by liquid peroxide should be thoroughly soaked by a hose stream. If their clothing absorbs peroxide, the corrosive action of the liquid will be accentuated by continual direct contact with the skin. They are now wearing a highly oxygenated uniform. Therefore, they must undress and wash down immediately. Severe skin reactions are possible from exposures.

Taken internally, hydrogen peroxide is poisonous. It is extremely dangerous to the eyes. In the event of these types of contact, or if skin blistering occurs, or skin irritation is not corrected by flushing with water, get medical attention at once. Recommended first aid measures, after dilution with water, include the use of boric acid paste on flesh burns and a 3 percent solution of boric acid in the eyes.

Self-contained breathing apparatus, of course, protects against some of these possibilities. Given the nature of hydrogen peroxide, flameproofed and nonabsorbent clothing is desirable. Workers around peroxide generally wear plastic aprons, gloves, and goggles. A safety shower is close at hand.

Peroxide Fire and Explosion Hazards

Hydrogen peroxide did not seem very dangerous during the discussion of the reactions in a 3 percent solution, a lather of oxygen on a cut, little streams of oxygen bubbles in a bottle, and an increased heat of decomposition coming from contact with a hand—heat so easily absorbed by the large percentage of water. Industry puts peroxide to work as a bleach or a polymerization promoter, or in the manufacture of rayons, dyes, starches, glues, and gelatins. When working with storage concentrations of 35 and 50 percent; with shipments of 70 percent solutions that are diluted to usable strengths at the scene of action; or even, for special purposes, with 90 percent strengths or pure H_2O_2, quiet little reactions become as strong and dangerous as the peroxide itself. Several interwoven hazards must be countered by special storage and handling requirements.

Hydrogen peroxide can be a more powerful oxidizer than chlorine or pure oxygen. Like most oxidizing agents, it will increase the burning rate of an involved flammable. When in contact with a combustible material (and a liquid can get very close indeed), hydrogen peroxide can lower ignition temperature or even cause a flammable material to ignite spontaneously. Explosive mixtures become possible as percentages become higher.

Explosive mixtures may be formed with many organic materials, but with rare exceptions, an initial hydrogen peroxide concentration of more than 55 percent is needed. The shock sensitivity and energy is then of the same order as TNT. Explosions can be caused by mechanical shock, heat or electrical discharge.*

*Factory Mutual, *Handbook of Industrial Loss Prevention,* 2nd ed. (New York: McGraw-Hill, 1967).

Commercial hydrogen peroxide will decompose into oxygen and water vapor, producing heat. However, this no longer involves a small bottle. Industrial shipping containers and storage tanks may hold thousands of gallons. Because it is possible for one volume of decomposing liquid to yield a thousand volumes or more of gas, an explosion is inevitable from the force of gas pressure alone, unless the container has the proper emergency venting devices.

The heat production that was so trifling in a 3 percent solution is now something else. Up to a certain percentage, there is still enough water in a peroxide solution to absorb the heat of decomposition without boiling. However, at a concentration of 64.7 percent or above, this is no longer true. Although water itself is a decomposition product and more water is continually made by the decomposition process, the percentage of H_2O_2 begins to rise as the water boils away. (Hydrogen peroxide boils at a higher temperature than water: 286°F [141°C].) The process becomes cumulative. The very heat it is creating triggers the peroxide into an even faster reaction, almost a definition of an unstable material. Eventually, no liquid at all will be left. In a 90 percent concentration, the heat of the decomposition gases is over 1,300°F (704°C); in a 100 percent concentration, their temperature is close to 1,800°F (982°C), more than enough to ignite any combustible around, without considering the presence of oxygen.*

Heat is not the only cause of hydrogen peroxide's decomposition. Just as the blood catalase did, many contaminants will greatly increase the rate of decomposition—for instance, dust and dirt, or such metals as brass, bronze, copper, iron, lead, manganese, silver, and their salts. Notice that many common structural metals are included in this list. Permanganates and hypochlorites are often specifically mentioned as being very dangerous in this respect. There is still another possibility: Although hydrogen peroxide is not flammable, a vapor concentration exceeding 26 percent in air is capable of violent decomposition, given a boost from a source of energy. This vapor–air mixture exists naturally above an enclosed concentration of 74 percent hydrogen peroxide at normal temperatures and pressures. It can also exist above lower concentrations when temperatures and pressures are increased.

At concentrations of 90 percent and more, hydrogen peroxide becomes a heavy, syrupy liquid and a firefighter's nightmare. The toxic hazard is extreme. Solutions, vapors, and mists are highly irritating. Explosive decomposition or explosive oxidation of combustibles at *ordinary temperatures* is possible. Above 200°F (93°C) an explosive decomposition of a 90 percent concentration is *to be expected.* Even if a mixture

---

*These temperatures are calculated assuming no heat loss to the environment. Such a theoretical situation is called **adiabatic.**

does not explode upon contact with its surroundings, the combination may subsequently detonate if it is shocked or catches fire. As discussed earlier, the heat of decomposition is extremely high. Decomposition can be caused not only by contaminants but also by light, temperature increase, agitation, or contact with rough surfaces. These liquids are workable at room temperatures only because they are handled gingerly and all contaminants and light are kept away. When used as rocket propellants, they are deliberately catalyzed into decomposition. The German V-1 used a hydrogen peroxide propellent during World War II, and the space program is still one of its major users.

Shipping

What can be done to control such a demon? The answer starts with stringent shipping requirements. Special containers are always required for hydrogen peroxide solutions. Amber or dark blue glass bottles are only two of a number of containers that are especially designed for peroxide. DOT shipping regulations show careful consideration of the increasing hazards of higher concentrations and a full awareness that iron and steel are on the forbidden list of contaminants. The following summary is drawn from Section 173.266 of the regulations. (For a complete description, refer to the original DOT regulations.)

Hydrogen peroxide solutions in water must be packed in containers as follows:

Over 52 percent

1. Glass bottles, not over one-quart (one-liter) capacity, packed inside two metal containers, one vented at the top, one at the bottom. Cushioning material, noncombustible, at least 10 times the volume of the liquid, wet with at least 10 percent water and a stabilizing agent.
2. Aluminum drums with sealed and vented closures. Venting arrangements as approved by the Bureau of Explosives. There should be signs reading KEEP THIS END UP or KEEP PLUG UP TO PREVENT SPILLAGE.
3. Tank cars must have venting arrangements as approved by the Bureau of Explosives.

52 percent or Less

1. Glass or earthenware containers of not more than one-gallon (four-liter) capacity, inside wooden boxes well cushioned with noncombustible material.
2. Carboys, single-trip drums, or 55-gallon drums, made of aluminum with vented closures.

Vented polyethylene and glass carboys. Vented aluminum bottles not over 5 pounds or pints. Polyethylene or other plastic bottles not exceeding one pint capacity inside wooden boxes.

In addition to the above, concentrations between 8 and 10 percent may also be shipped in paraffin-lined wooden barrels.

Small amounts of hydrogen peroxide, 52 percent or less, are exempt from the above but must be shipped in glass bottles inside metal cans. There are no exemptions when the concentration is over 52 percent.

Storage

Caution must be maintained after these containers reach their destination. Because drums are vented, they should not be stacked. (This also prevents a damaging fall.) Store and handle them in an upright position. To prevent contamination, store peroxides in the original containers, kept closed against contaminants or water escape that will increase the concentration of the remaining solution. Simply, store drums in a well-ventilated area where contaminants and sources of heat cannot disturb them. A detached, noncombustible building is ideal, although drums are often kept outside under a protective nonflammable canopy. Storage distances from occupied or important buildings depend again upon the strength of the solution. Below 64.7 percent, drums may be stored within 50 feet. At or above this percentage, 100 feet (30 meters) is the minimum. Limited storage of concentrations below 64.7 percent is allowed inside a main building if the storage room has walls of fire-resistive construction, nonflammable floors, drainage facilities, and adequate supplies of water for dilution of spills.

Special alloys of aluminum are used to build storage tanks. Piping is designed to prevent a backflow of contaminants or a trapping of peroxide. Even the gaskets are made of an inert material that will not catalyze H_2O_2. Dilution techniques and location of unloading areas are subject to strict regulations. In spite of all this, and dust filters too, decomposition can still start inside a large tank of highly concentrated peroxide. If it does, the importance of a separated location and adequate pressure-relief devices on tanks becomes evident. Tanks are located outside in a position dictated by their size, strength of the solution they contain, and how the pressure inside them is to be relieved. Pressure-relief devices are sized according to the tank they must vent, and they should be large enough to accommodate a severely contaminated solution. These devices include large manhole covers that vent at one psi, and rupture discs that let go at 15 psi (1 atmosphere). Equipped with manhole covers, tanks containing concentrations below 64.7 percent can be stored next to a noncombustible wall or 25 feet (7.6 meters) from a combustible wall. Approximate storage distances from an important building for a 2,000-gallon (7.54-cubic-meter) peroxide tank, with and without pressure-relief devices, are given in Table 11–3.

TABLE 11–3. Storage Distance for a Hydrogen Peroxide Tank of 2,000 Gallons (7.54 M³)

Percentage of Solution	Distance [a]		Distance [b]	
	Feet	Meters	Feet	Meters
35%	40	12	500	152
50	80	24	600	183
70	150	46	700	213
90	300	91	850	259
100	600	183	1,200	366

[a]Distance for: 35% and 50% solutions equipped only with 15 psi (1 atmosphere) rupture discs; 70%, 90%, and 100% solutions with manhole covers.
[b]Distance for: tanks not equipped with pressure relief. If tanks are barricaded, this distance can be cut in half. A tank farm should be located according to the distance required for largest tank.

Firefighting

When one is called into an emergency involving hydrogen peroxide, there is an immediate need for information—the correct answers to a number of questions. There may never be a time when technical advice is needed more. The first question is, What has happened? Hydrogen peroxide may have spilled, formed an accidental mixture with a combustible, be exposed to fire, or the solution inside a tank or smaller container may be decomposing. Once you know what has happened, you can begin to think about a succession of further questions.

If there is a spill, how much water will be necessary to dilute it to a safe level? Where will all this liquid flow? Will people be exposed to a toxic hazard? What is the possibility of contamination? If the spill is continuing from piping, where are the remote shutoffs?

If a tank or a group of containers is threatened by, or involved in, a fire, what is its location in relation to important exposures? How are the tanks vented? Will the pressure-relief devices be able to take care of the overpressure? Are the tanks aluminum rather than steel, the containers aluminum or plastic? Is it possible to keep them intact with a cooling spray stream?

What can be done about decomposition, particularly if a tank contains a concentration above 64.7 percent? Because emergency stabilization may be possible, who would know what stabilizers are needed and how to use them? Once again, are the containers properly vented, or is there a possibility of a pressure explosion? If the temperature inside a tank is 30 degrees above normal, or rising at the rate of one degree every 15 minutes, it probably means that the situation within is deteriorating. Does the plant have a system that automatically monitors temperatures and sounds an alarm? Can this system also dump a tank into a safely diked location while simultaneously diluting the contents? Armed with answers to these questions and to the others that will occur at the scene, fire officials can put together some sort of workable plan.

To help, a strong supply of water must be available. Nothing can replace it to cool tanks and containers, to dilute and wash away spills, to protect exposures. With an oxidizer involved, it is the only efficient extinguishing agent available. However, if this fails to improve the situation, evacuation may become necessary. The distance to withdraw depends upon an estimate of the situation, once again after consultation with plant personnel.

INORGANIC PEROXIDES

When one of these double "O" peroxy groups joins with a metal, other inorganic peroxides are formed. (See Figure 11–2). All are nonflammable oxidizing agents. The most common is **sodium peroxide**, Na_2O_2, a yellowish-white powder that becomes yellower as its temperature rises. Hard lumps of this peroxide, called oxone, are sometimes encountered.

Sodium peroxide is quite versatile. It can be used to bleach paper, textiles, fats, oils, resins, animal products, bristles, straw, ivory, sponges, and feathers. It is also employed as a germicide, a deodorant, an oxidizing agent, in dyes, and for water purification. One of its more interesting functions is to regenerate the air in submarines, aircraft, and space vehicles. Sodium peroxide will combine with carbon monoxide to form harmless soda ash, sodium carbonate.

$$Na_2O_2 \ + \ CO \longrightarrow Na_2CO_3$$

Sodium   Carbon   Sodium
peroxide   monoxide carbonate

Sodium peroxide will also react with the carbon dioxide in the air to form soda ash and fresh oxygen.

$$Na_2O_2 \ + \ 2CO_2 \longrightarrow 2Na_2CO_3 \ + \ O_2$$

Sodium   Carbon   Sodium   Oxygen
peroxide   dioxide   carbonate

(A reaction like this, using potassium superoxide, KO_2, is put to work in Chemox and other oxygen-generating gas masks.)

Inorganic peroxides, built around one of the **alkaline earth** metals (magnesium, calcium, strontium, and barium), are considered to be less water-reactive than those built around the **alkali** metals (lithium, potassium, and sodium). Sodium peroxide reacts vigorously with water. Depending upon the ratio of water to peroxide, the reaction can become explosively rapid, releasing oxygen and a great deal of heat. The heat of reaction, especially in the presence of oxygen, is more than enough to ignite most nearby combustibles. A mixture of sodium peroxide and a flammable material can either be explosive or easily ignited by a relatively small source of heat, even by friction. The rate of combustion is very high. These peroxides must be kept away from flammable liquids; immediate ignition may take place, especially in the presence of water.

Sodium peroxide is shipped in a wide variety of containers: glass bottles, fiber cans or boxes, wooden kegs, boxes or barrels, ranging in weight up to 400 pounds (182 kg). Some of these containers are combustible; some of them can be broken. For obvious reasons, spilled peroxide must be cleaned up promptly. In storage, it should be separated from flammable materials and protected from damage or moisture.

Firefighting procedures depend upon circumstances. A very small fire involving alkali metal peroxides can be isolated by dry chemical or even drowned if enough water is used. The heat of reaction is absorbed by the flooding. When larger amounts of peroxide are involved, water simply cannot be used on the material itself but can wet down exposures and limit the spread of the fire.

Finally, sodium peroxide is caustic itself and will react with wet skin to form heat and a corrosive alkali; either can result in a serious burn. Avoid getting it on the skin or in the eyes, or breathing the dust. Wear full protective clothing. See Table 11–4 for the toxic and reactive properties of other peroxides.

NITRATES

Sodium Nitrate

Continuing our practice of presenting chemistry in digestible amounts, a short discussion of how an acid reacts with a base to form a salt is somewhat revealing. When this acid is **nitric**, the salt formed is called **nitrate**.* Economically, nitrates are the most important of the oxidizing agents. The most widely used of them is one variously called caliche, soda niter, Chilean saltpeter, or sodium nitrate. The names reflect the fact that it is not only produced synthetically but also mined in several areas of the world. It is easy to understand why tons of it can be piled up in a single warehouse when looking over a list of its uses: to manufacture other nitrates, sulfuric and nitric acids, glass, pyrotechnics, medicines, matches, some forms of dynamite and military explosives, dyes, and food preservatives; or as a fertilizer, an oxidizing material, a brine refrigerant, and one of the salts in molten baths.

---

* An acid ending in -ic always forms a salt ending in -ate; -ous acids form -ite salts: nitric to nitrate, nitrous to nitrite, and so on.

TABLE 11–4. Inorganic Peroxides

	Properties	Shipping and Uses
Barium peroxide BaO_2 (Barium dioxide; Barium superoxide)	Grayish-white powder. *Poisonous.* Avoid breathing dust and contact with skin. Decomposes slowly in cold water, more rapidly in hot water. Above 1,100°F (593°C) it will decompose into oxygen and barium oxide, which is caustic and also poisonous. Melts at 842°F (450°C). Oxidizing agent.	Shipped in bottles, boxes, kegs, drums, barrels, multi-walled paper sacks. Uses: manufacture of oxygen and hydrogen peroxide, glass decolorizer, bleach, tracer bullets, to oxygenate water.
Calcium peroxide CaO_2	White or yellowish, odorless powder. Decomposes at 527°F (275°C). Practically insoluble in water. Oxidizing agent.	Shipped in drums up to 200 pounds (91 kg.). Uses: seed disinfectant, dentifrices, medicine.
Lithium peroxide Li_2O_2	White powder or yellow grains. Decomposes in water, giving off oxygen and heat. Water solutions at ordinary temperatures can be decomposed.	Uses: bleach, source of oxygen.
Magnesium peroxide MgO_2	White odorless powder. Gradually decomposed by water with release of oxygen. Oxidizing agent.	Shipped in drums up to 200 pounds (91 kg.). Uses: bleach, oxidizer, medicine.
Potassium peroxide K_2O_2	Yellow mass. Decomposes in water, giving off oxygen and quantities of heat. *Caustic.* Does not burn or explode alone, but when mixed with combustibles, these mixtures are explosive or ignite easily. See discussion on sodium peroxide in text.	Shipped in glass bottles, cans, boxes, kegs, barrels, metal drums. Uses: oxidizing agent, bleach, oxygen-generating masks such as the Chemox.
Strontium peroxide SrO_2 (Strontium dioxide)	White odorless powder. Decomposes in hot water or when heated. Mixtures with combustibles ignite easily or can explode. Oxidizing agent.	Shipped in drums up to 500 pounds (227 kg.). Uses: bleach, medicine, fireworks.

With the exception of ammonium nitrate and nitrite, cases unto themselves, most of the nitrate and nitrite salts have similar properties, varying only in some details. Table 11–5 lists many of the nitrate salts.

(Although sodium nitrite looks a little different, having slightly yellowish crystals instead of the pure white or colorless crystals of sodium nitrate, it is shipped under the same DOT yellow label and subject to the

TABLE 11–5. Properties of Nitrates

	Properties	Uses
Barium nitrate (Nitrobarite) $Ba(NO_3)_2$	Lustrous white crystals. *Poisonous.* May be lethal if swallowed. Melts at 1,097°F (592°C).	Pyrotechnics (gives a green flame), explosives.
Calcium nitrate (Norway saltpeter) $Ca(NO_3)_2$	Colorless granules. Evolves heat when it dissolves in water. Decomposes at 270°F (132°C). Melts between 44°F (7°C) and 1,040°F (560°C), depending upon amount of water in the molecule.	Explosives, pyrotechnics, matches, radio tubes.
Cobaltous nitrate $Co(NO_3)_2$	Red crystals. Melts at 133°F (56°C). Decomposes at 165°F (74°C).	Inks, cobalt compounds and pigments, decorating stoneware.
Copper nitrate $Cu(NO_3)_2$	Blue crystals. *Poisonous.* Melts in range between 79°F (26°C) and 238°F (114°C). Decomposes at 338°F (170°C).	Medicine, preparation of light-sensitive papers, dyes.
Lead nitrate $Pb(NO_3)_2$	White crystals. *Poisonous.* Decomposes at 878°F (470°C).	Lead salts, medicines, paints, matches, special explosives.
Lithium nitrate $LiNO_3$	Colorless powder. Melts at 491°F (255°C).	Low-temperature salt baths, heat-exchange medium, ceramics.
Magnesium nitrate $Mg(NO_3)_2$	White powder. Melts at 203°F (95°C). Decomposes at 626°F (136°C).	Pyrotechnics.
Nickel Nitrate $Ni(NO_3)_2$	Green crystals. Melts at 133°F (56°C). Boils at 277°F (136°C).	Nickel plating, brown ceramic colors, nickel catalysts.
Potassium nitrate (Saltpeter, niter) KNO_3	White crystals or powder. Melts at 631°F (333°C). Decomposes at 752°F (400°C).	Pyrotechnics, salt baths, pickling meat, glass manufacture, treating tobacco to make it burn evenly.
Silver nitrate $AgNO_3$	Colorless to white, large or small crystals, which become grayish to black on exposure to light, or in the presence of organic materials. *Poisonous and corrosive.* May cause burns, dangerous to eyes. Wear protective equipment. Melting point at 414°F (212°C). Decomposes at 831°F (444°C).	Photography, hair dyeing, silver plating, silvering of mirrors, indelible inks, silver salts, medicine. Shipped in amber bottles up to 200 ounces (5.7 kg).

continued

TABLE 11–5. *Continued*

	Properties	Uses
Strontium nitrate $Sr(NO_3)_2$	White crystals. Melts at 1,058°F (570°C).	Marine signals, matches, railroad flares (red fire).
Thorium nitrate $Th(NO_3)_4$	White crystals or solid. Decomposes at 932°F (500°C).	Medicine, fluorine detection.
Uranyl nitrate (Uranium nitrate) $UO_2(NO_3)_2$	Yellow crystals. *Mildly radioactive;* protect personnel. *Dust can be toxic.* In ether solution, do not allow to stand in sunlight. Melts at 141°F (61°C). Boils at 244°F (118°C).	Photography, uranium glazes, medicine, uranium extraction.
Zinc nitrate $Zn(NO_3)_2$	Colorless lumps or crystals. Melts at 97°F (36°C). Boils at 268°F (131°C).	Medicine, dyeing.

same storage regulations. In any event, it slowly oxidizes to sodium nitrate when exposed to air. After a time, its properties will be identical.)

What makes sodium nitrate troublesome is that it is an oxidizer, it is hygroscopic, it decomposes, and it melts. These four phrases summarize its properties.

Sodium nitrate is an oxidizing agent. When heated, it gives off oxygen. What will happen when it meets a flammable material depends on a large number of variables: How much of each is involved? What kind of flammable? How complete is the mixture? How hot is it? Violently rapid combustion is possible. So is an explosion. Certainly, the degree of flammability will be increased, and ignition will be made easier. However, nitrate-flammable mixtures generally require an ignition source before they burn or explode. This is not always the case when a chlorate or a high concentration of hydrogen peroxide is mixed with a combustible.

In any event, why risk finding out if an ignition source is needed? Far better to have all nitrates stored in a cool, dry place, entirely separated from combustibles. However, many buildings have wooden floors or favor storage on wood pallets. Sodium nitrate, being hygroscopic, will absorb moisture either from the air or from materials it contacts. This property has an interesting effect on wood. The nitrate dries the wood, making it crack and curl up into a series of tiny splinters. If these wood splinters ignite, fires of amazing intensity and rapid spread are possible because they are fanned by the presence of an oxidizing agent. Splintering is not confined to floors or pallets. Nitrate dust, settling on wooden ledges or beams, will convert all wooden exposures into several million wooden slivers, all awaiting ignition.* Even if the building is noncombustible, sodi-

---

*Other nitrates may be less hygroscopic. Potassium nitrate is one example. Plant protection hose lines can become defective when exposed to a dusty nitrate atmosphere for a time. Consider this during inspections.

um nitrate is shipped in a wide variety of flammable containers, ranging from fiber drums to multi-walled paper bags that may or may not be moisture-proofed. If the containers become wet, the nitrate may partially melt and impregnate them. Upon drying, the containers become highly flammable. For all these reasons, noncombustible storage bins are recommended for bulk nitrate shipments. Shipping bags and barrels are supposed to be thoroughly washed or destroyed after use. Unfortunately, this safety precaution is usually not observed.

Like other nitrates, sodium nitrate will decompose. When decomposition starts, at 716°F (380°C), sodium nitrate begins to release the same insidious, delayed-action group of colorless to tan to orange to brownish gases as cellulose nitrate, the nitrogen oxides. These gases will poison firefighters before they are aware of it. Firefighters must protect themselves in any nitrate fire with a self-contained breathing apparatus. Table 11–5 gives the decomposition temperatures of other nitrates.

Sodium nitrate melts at 586°F (308°C), well within ordinary fire temperature range. A molten oxidizing agent, loose in a fire situation, is something to be feared. It can easily ignite combustibles as it runs around with its load of oxygen, looking for trouble. If we pour water on it, or if water and molten salt flow about until they meet, steam explosions will occur. The intensity of the fire, fed as it may be by finely divided splinters of wood and generous supplies of oxygen, is overwhelming. No wonder sodium nitrate fires have been so destructive; no wonder a sprinkler system occasionally melts.

Quick detection and action are all-important in nitrate fires. Without a fast knockdown, extinguishment may be impossible. Watchmen or automatic alarms are essential, along with a sprinkler system and a fire department that can apply enough water to put out the fire in a hurry before molten salt forms. Once a nitrate melts in large quantities, using water (at close quarters) will be dangerous, and no other extinguisher is going to be much better.

Salt Baths

Salt baths come in three temperature ranges with each range particularly suited for heat-treating a specific metal. Various chlorides are used in high-temperature baths, between 1,100°F and 2,450°F (593°C and 1,343°C); cyanide salts are found in medium-temperature baths between 1,000°F and 1,750°F (538°C and 954°C). We are concerned here with the so-called **low-temperature baths** that operate in a range between 500°F and 1,100°F (260°C and 593°C). These baths often contain molten nitrates, such as a combination of potassium nitrate and sodium nitrate.* Explosions are not uncommon in salt baths for several reasons: overheating, reaction with surroundings, reaction with metals, and trapped water or air.

---

*Although sodium nitrate melts at 586°F (308°C), and potassium nitrate at 631°F (333°C), a mixture of the two melts at 426°F (219°C).

Overheating. Whether because of failure of temperature controls or because of a mistaken setting, a salt bath can overheat. Nitrates start to decompose at temperatures around 750°F (399°C). For a while, the decomposition moves along slowly. At 1,100°F (593°C) it speeds up, releasing quantities of the nitrogen oxide gases. Above 1,300°F (704°C), violent decomposition is possible. Normally, operating temperatures of nitrate baths should be below 1,000°F to 1,100°F (538°C to 593°C).

Reaction with Surroundings. These molten salts are oxidizing agents in a most dangerous condition. Careless storage of reserve stocks of nitrate, chemical reactions with a combustible within the bath, or even a reaction with the metal of the container itself are frequent causes of trouble. And, if the nitrate gets loose, the consequences can be disastrous.

One fire report speaks of the "molten nitrate scattered about after an explosion, setting fire to surrounding combustibles." Another report mentions how a welded seam was ruptured, "which allowed the molten nitrate to react violently with the accumulated carbon in the furnace." After another explosion, an analysis of the salt residue "indicated that one can of cyanide had been mistakenly added with the nitrates to the nitrate bath. The combination of the salts caused the explosion as the temperature was increased." Even this is possible: "sections of the steel plate of the salt bath container were reduced from 1/2" to 1/8" by the rapid oxidation of the salts."

Reaction with Metals. It has already been discussed how greedy magnesium is for oxygen. Imagine what happens if magnesium is introduced into a nitrate salt bath that can supply all the oxygen it hungers for. "An explosion and fire occurred in a plant which was heat treating aluminum alloys. Magnesium alloy castings were accidentally dipped into the nitrate bath which was heated to a temperature of approximately 1,000°F. [538°C.]. An explosive reaction followed and the plant and equipment were almost completely demolished."

Aluminum itself has caused explosions, either in overheated baths or by a thermite reaction, caused when particles of aluminum react with the iron oxide sludge on the bottom of the salt container.

Trapped Water or Air. Salt baths have exploded because of trapped air or because water has been brought into the situation. Steam explosions have taken place when water comes into contact with the molten salt "either as a carry-over from a preliminary cleaning bath, [or from] overheated service piping, leaky roofs, operation of automatic sprinklers, and liquid foods placed on ledges near the baths for warming."

As firefighters, consider the following case:

A fire occurred in the heat treating section of a large metal working plant which resulted in an explosion. The fire department was partly responsible for this explosion by carelessly deluging the heat treating section with streams of water. The water caused a tremendous steam explosion when it suddenly entered one of the molten salt baths. The steam pressure created was sufficient to destroy the structure.

Consequently, the use of sprinklers around nitrate salt baths has had considerable scrutiny. The results of tests were summarized this way:

> Sprinkler discharge at different pressures was directed onto typical molten salt baths without explosive results and without spattering of any significance from the standpoint of personal injury and property damage. In the absence of sprinklers, a serious fire must ultimately be fought with hose streams; there is record of personnel being driven from the building by spattering when a solid hose stream was accidentally directed into a salt bath. Hose streams cause more severe spattering than does the relatively gentle discharge from sprinklers. A non-combustible hood may be installed over the baths to prevent sprinkler water from contacting them directly.*

"Spattering" in the preceding quotation equals "steam explosion."

Emergency Procedures. When called in because of an overheated salt bath, make sure the heat supply to the furnace has been shut off and that all work has been removed from the bath. If the temperature is still rising, get ready for an explosion.

> Fires usually result from spillage and leaks. They spread fast and are difficult to control. Dry sand, which should be stored in nearby buckets and bins, can confine and prevent the spread of the escaped melt. Carbon dioxide and approved dry powder type extinguishers may be used to extinguish burning carbonaceous material around the immediate vicinity of the salt bath. No vaporizing liquid (carbon tetrachloride), water, foam, or other aqueous extinguishing agents should be permitted in fighting molten salt fires. While it is obvious that the addition of water to a molten salt bath will in all probability cause formation of dangerous steam pockets and an explosion, precautionary warning notices (CHEMICAL BATH—USE NO WATER!—EXPLOSION DANGER!) should be placed in prominent locations to prevent mistakes. All persons in the neighborhood of baths should be instructed as to the dangerous practice of using water and the correct use of sand.
> Cautioning placards to warn firemen outside the factory of the presence of nitrate baths should be displayed at the entrances of the plant. The management should invite the officers of the nearby fire companies to inspect and become thoroughly acquainted with the location of these baths.

One last word of caution: Besides the obvious fact that firefighters will be dealing with salts hot enough to destroy flesh, do not forget the danger from breathing the nitrogen oxides. If a uniform somehow becomes impregnated with a nitrate salt, remember that these are oxidizing agents. If clothing catches fire, a severe, perhaps fatal, burn can occur. Change clothes at once.

SUMMARY

Although oxidizers are relatively nonflammable, their discussion in this text is necessary because of their violent reactivity to organic materials. Of particular concern to firefighters are the organic peroxides that are

---

* Factory Mutual, *Handbook of Industrial Loss Prevention,* 2nd ed. (New York: McGraw Hill, 1967).

already contaminated with an organic material. These compounds are kept stable either chemically or through refrigeration. Some require heating in order to be industrially utilized. Most incidents involving compounds of this type occur as the result of equipment or human error. Additionally, oxidizers contain oxygen; when involved in a fire, they contribute significantly to the severity of the fire by providing increased levels of oxygen. Basically, an oxidizer is capable of providing all four sides of the fire tetrahedron, by itself.

REVIEW QUESTIONS

1. Chlorites, chlorates, and perchlorates are capable of exploding during a fire. Name three ways this can happen.
2. What procedures may be necessary if ammonium perchlorate is threatened by a large fire?
3. How can the decomposition rate of hydrogen peroxide be increased?
4. At what percentage is a DOT yellow label, oxidizer, required on hydrogen peroxide solutions?
5. What percentage of hydrogen peroxide is generally required to form an explosive mixture with organic materials?
6. Why are concentrations of hydrogen peroxide of 64.7 percent and above so dangerous?
7. Name five materials that will greatly increase the decomposition rate of H_2O_2.
8. What reactions occur when sodium peroxide is contacted by water?
9. Sodium nitrate is hygroscopic. What does this mean, and why is it important to firefighters?
10. What can happen when sodium nitrate melts in a fire?
11. Discuss the use of hose streams around molten salt baths.
12. Discuss the use of a sprinkler system around molten salt baths.
13. Why should empty bags and barrels that contained sodium nitrate be washed or destroyed after use?
14. Why is stacking hydrogen peroxide containers in storage a bad practice?
15. Discuss the first aid measures for a firefighter splashed by hydrogen peroxide.

BIBLIOGRAPHY

1. *Handbook of Industrial Loss Prevention*, 2nd ed., Factory Mutual System, 1967.
2. "Molten Salt Bath Furnaces," Standard No. 86C, Chapter 14, National Fire Protection Association, 1 Batterymarch Park, Quincy, MA.

FURTHER READINGS

Medard, Louis A. *Accidental Explosions Vol. I: Physical and Chemical Properties,* Englewood Cliffs, NJ: Prentice Hall, 1989.
Medard, Louis A., *Accidental Explosions Vol. II: Types of Explosive Substances,* Englewood Cliffs, NJ: Prentice Hall, 1989.

VISUAL AIDS

A color film, "Chemical Booby Traps," produced by General Electric, has a section on the effect of oxidizers on combustible materials. There are also two films for a class interested in the chemistry of the situation: "Acids, Bases, and Salts," 20 minutes, color, and "Oxidation and Reduction," 10 minutes, color. These films can be obtained from a film library that is oriented toward college-level scientific visual aids. Your local college can assist you.

Oxidizers, Identification, Properties, and Safe Handling, Hazardous Materials Library, FEMA Region IX, Presidio of San Francisco, FL5-6-402. (54 minutes, VHS format)

Reactive and Explosive Materials, Hazardous Materials Library, FEMA Region IX, Presidio of San Francisco, FL5-6-234A. (60 minutes, VHS format)

DEMONSTRATIONS

Fire from Ice: Both the water reactivity of sodium peroxide and the accelerated burning rate given a combustible by an oxidizing agent can be demonstrated rather dramatically. You will need some excelsior or shredded paper, a small amount of sodium peroxide, an ice cube, and a noncombustible container. You can make quite a production of this demonstration. Work outdoors. Build a bed of excelsior on the bottom of your container. Sprinkle a couple of tablespoons of sodium peroxide on the excelsior. Put the ice cube on top of the peroxide. The cold water slows down the reaction enough to allow a getaway. Within a minute or less, enough water has melted to activate the peroxide, which will then ignite the paper through the heat of reaction. Keep the class upwind. Sometimes the paper ignites with a pop.

The reactions in a bottle of household peroxide, described in the text, can easily be demonstrated.

Afterword

They say the hardest sentences to write in a book are the first and the last. I can thank James Meidl for the first and this final incident for graphically summarizing the entire intent and purpose for writing this book. This illustrative incident was unusual in that it contains a majority of the hazardous properties of flammable hazardous materials in a single incident. The following report is taken from a National Transportation Safety Board, Hazardous Materials Accident Spill Maps report #NTSB-HZM MAP-80-4.

About 4:22 P.M. on September 8, 1979, two locomotives and cars in train positions 1 through 33 of a 53-car train derailed about three miles east of Paxton, Texas. Sixteen derailed cars contained DOT-regulated materials including acetaldehyde (Nos. 19, 20, 21, and 26), butadiene (Nos. 7, 30, 31, and 33), ethyl acrylate (No. 18), ethylene oxide (No. 14), hydrogen fluoride (No. 9), isobutylene, LPG (No. 6), methyl alcohol (No. 17), tetrahydrofuran (No. 8), and vinyl acetate (Nos. 15 and 32). Other derailed cars contained rubber (Nos. 1, 2, 3, 4, 5, and 25), plastic pellets (Nos. 22, 23, and 24), ethylene glycol (No. 16), propylene glycol (Nos. 10,11, and 12), and diabasic ester (No. 13). Cars in train positions 27, 28, and 29 were empty.

During the derailment, which lasted about 80 seconds, car No. 7 struck car No. 6, which severed car No. 7's top housing and damaged its fittings. Butadiene escaped from No. 7 and spread as a "bubbling" ground fog approximately 300 yards across the derailment area and nearby residential property. Cars Nos. 8 and 9 were not leaking. The bottom outlet valves of cars Nos. 11 and 12 were damaged as they rolled. Propylene glycol escaped through damaged lower outlet valves of cars Nos. 10, 11, and 12.

Cars Nos. 13 through 17 jackknifed. Cars Nos. 13 and 14 remained intact. Car No. 16 punctured car No. 15 at the head. Vinyl acetate escaped and spilled around car No. 14. The bottom outlet of car No. 16 was torn from the tank shell, allowing ethylene glycol to spill around cars Nos. 14 and 15. Car No. 17 received a small head puncture through which methanol escaped. The bottom outlet of car No. 18 was damaged, and ethyl acrylate escaped.

Acetaldehyde escaped through a breach in the shell of car No. 19. The bottom outlets of cars Nos. 20 and 21 were damaged, and acetaldehyde escaped. Cars Nos. 24 and 26 collided. The head of car No. 26 was punctured, and acetaldehyde escaped and pooled around cars Nos. 19 through 23. Cars Nos. 30, 31, 32, and 33 remained intact.

Witnesses saw the butadiene cloud ignite when cars Nos. 24 and 26 collided. As nearby residents fled, four received burns on their feet and four inhaled smoke. Spilled flammable liquids ignited and burned.

The Joaquin Volunteer Fire Department arrived on the scene by 4:30 P.M. The Joaquin VFD and the Shelby County Sheriff's Office evacuated between 200 and 300 people from a two-mile radius of the derailment site by 6:00 P.M.

The intense fire and smoke surrounding the derailed cars increased the internal temperature and internal pressure of intact cars Nos. 8, 9, and 14. At 5:30 P.M. and later in the night, witnesses heard explosions as the tetrahydrofuran and ethylene oxide tanks violently ruptured. Hydrogen fluoride escaped through the safety relief valve of tank No. 9.

Heat from the burning butadiene escaping from tank car No. 7 burned a hole in the shell of tank car No. 6. Isobutylene (LPG) escaped and ignited.

Witnesses estimated the fire that resulted from the released products and burning rubber and plastics to be about 500 feet high. A black smoke cloud evolved. Residents in Center observed a black smoke cloud overhead by 6:00 P.M.

During the night, the fire subsided yet continued to consume flammable and combustible matter. The Joaquin VFD monitored the blaze. For the next two days, small fires burned the remaining products. By the morning of September 11, the fires were completely extinguished. All injuries resulted from minor burns and smoke inhalation.

Index

Benzene, 39-41. *See also* Toluene fires, extinguishment
  flammable ranges of, 89 (table 5-4)
  poisoning, 39-41
  properties of, 40 (table 3-6)
Beryllium, 173 (table 9-1), 180-81
Beryllium powder, 181
Binders, plastics, 227
Bismuth, 173 (table 9-1), 194-95
Bituminous coal, 147
BLEVEs (Boiling Liquid Expanding Vapor Explosions), 69-70
Boiling points of flammable liquids, 38-39, 130-31 (table 7-1)
Bone black, 147
Boron, 197-98
Boyle's Law, 80-81
British Thermal Units (BTU), 4
Bulk flammable liquids. *See* Flammable liquids, bulk handling of
Butane, 33, 109-11
  boiling point of, 130 (table 7-1)
  flame temperatures of, 102 (table 5-6)
  flammable ranges of, 89 (table 5-4)
  properties of, 34 (table 3-5)
  properties of liquefied, 111 (table 6-1)
  vapor pressure at normal temperature, 123 (fig. 6-1)
Butyl acetate, properties of, 57 (table 3-13)
Butyl alcohol, 44 (table 3-8)
  properties of, 46, (table 3-9)
Butylaldehyde, properties of, 58 (table 3-13)
Butyl amine, properties of, 55 (table 3-12)
Butyl ether, properties and uses of, 52 (table 3-11)

C

Cadmium, 195
Calcium, 173 (table 9-1), 181-82
Calories, 4
Camphor, 146, 151
  fire properties of, 145 (table 8-1)
  shipping and storage, 146
  use in plastics, 201, 211
Carbon, 146-51
  elemental, 146-47
  hazards of, 149-61
  organic compounds and, 146
  poisonous and explosive qualities of, 151
  spontaneous heating and ignition, 149-61
Carbon atom chains, 30, 33, 35
Carbon blacks, 148-49
  crystals, 148-49
  uses of, 148
Carbon dioxide, 151
  properties of pressurized, 84 (table 5-1)
Carbon disulfide, 161-66
  DOT labeling of, 144
  fire extinguishment of, 163
  hazards rating of, 161-63
  properties of, 31 (table 3-3), 157 (table 8-3), 161-63

storage and shipping, 163-64
  toxicity of, 162
  uses of, 162-63
Carbon monoxide, 9, 12, 151
  flame temperatures of, 117 (table 5-9)
Carbonyl group, 57, 60
Casein, 227-28
Catalyst, 45
Celluloid, 201
Cellulose nitrate, 201-8
  automatic sprinkler protection for, 207
  camphor content, 201
  DOT labeling of, 203 (table 10-1), 204, 205
  fire fighting, 207-8
  fire hazards of, 202-4
  gases formed by, 205 (table 10-2)
  nitration process, 202
  shipping, 205-7
  storage, 206-7
  toxicity of, 203-4
  types of, 203 (table 10-1)
Cellulose plastics, 228-29
Cetyl alcohol, 44
Channel black, 148
Charcoal, 146
Charle's Law, 81
Chemical
  bonding, 22-23
  chain reaction, 2
  equations, 25-26
  foaming of plastics, 249
  formulas, 24-25
  mixtures, 19
  process symbols, 25
  valence, 22
Chlorates, chlorites, 244 (table 11-1)
Chlorine
  vapor pressure at normal temperature, 123 (fig. 6-1)
Chromium, 173 (table 9-1), 191-92
Coal, 146-47
Cobalt, 192
Cold molded plastics, 230
Colors of gas containers, identifying, 87 (table 5-2)
Combined Gas Law, 81-83
Combustible liquids. *See also* Flammable liquids, classes
  definition of, 30
Combustible metals, 169-99
  aluminum, 173 (table 9-1), 189-90
  antimony, 173 (table 9-1), 194
  barium, 173 (table 9-1), 181-82
  beryllium, 173 (table 9-1), 180-81
  bismuth, 173 (table 9-1), 194-95
  cadmium, 195
  calcium, 173 (table 9-1), 181-82
  chromium, 173 (table 9-1), 191-92
  cobalt, 192
  combustibility conditions: environments, 172, 175; form and shape, 170-172; oxidation resistance, degree of, 170

definition of, 241
group classifications of, 242
hazards of, 241-42, 243-44
hydrogen peroxide, 246-54; as antiseptic, 248; fire and explosion hazards of, 249-51; firefighting, 253-54; as rocket propellants, 251; shipping requirements, 251; storage recommendations, 252-53; tanks, 2000-gallon, distances for, 253 (table 11-3); toxicity of, 248-49
identification of, 242-43
inorganic peroxides, 242, 273 (fig. 11-2), 254-55
nitrates, 255-61; automatic sprinklers and, 259-61; properties of, 257-58 (table 11-5); salt baths, 259-61
per- compounds, 246
permanganate, persulfate, perborate group, 242, 245 (table 11-2)
peroxides, 246
sodium nitrate, 255-59; fire-fighting procedures for, 259; flammability of, 258-59; salt baths, 259-61
sodium peroxide, 254-55; combustibility, 254; reactive properties, 254; toxicity, 255; uses for, 254
toxic and reactive properties of, 256 (table 11-4)
Oxygen, 8, 269
and combustion, 11-12
liquid, 132 (table 7-2), 137-38
and open flames, 9
properties of cryogenic, 132 (table 7-2)
properties of pressurized, 85 (table 5-1)
and spontaneous ignition, 9

P
Paraffin series, 33-39
properties of, 34 (table 3-5)
Paraldehyde, properties of, 58 (table 3-13)
Pentane, 35
boiling point of, 131 (table 7-1)
flammable range, 89 (table 5-4)
properties, 34 (table 3-5)
Perborates, 243, 245 (table 11-2)
Per- chemical compounds, 246
Perchlorates, 243, 244-46, 244 (table 11-1)
Permanganates, 243, 245 (table 11-2)
Peroxides, 246
persulfates, 242, 245 (table 11-2)
Petroleum ether, 37
fires, extinguishment of, 38
flammable ranges of, 89 (table 5-4)
Petroleum fractions, properties of, 32 (table 3-4)
Phenolics, 232-33
Phosphorus, 151-55
black, 151
bombs, 152
compounds of, 154-55
DOT labeling of, 152 (table 8-2)
ignition temperature, 151
properties, 152 (table 8-2)

red, 154
white, 151-54
Phosphorus compounds, 154-55
Phosphorus pentasulfide, 155
Phosphorus sequisulfide, 154-55
Plastic textiles, 229-30
Plasticizer, 202, 211
Plastics, 201-38
ABS plastics, 226
acetyl resin, 226
acrylics, 226-27
amino plastics, 227
binders, 211
burning rate groups, 219
burning rates of plastic films, 221 (table 10-5)
casein, 227-28
celluloid, 201
cellulose nitrate (nitro cellulose); automatic sprinkler protection for, 207; camphor content, 201; DOT labeling, 203 (table 10-1), 204-5; fire fighting, 207-8; fire hazards, 202-4; gases formed by, 205 (table 10-2); nitration process, 202; shipping, 205-7; storage, 206-7; toxicity, 203-4; types of, 203 (table 10-1)
cellulose plastics, 228-29
chemical foaming of, 220-21, 223
cold molded, 230
coumarone-indene resins, 230
definition, 201
epoxies, 230-31
explosive properties of plastic dust, 222-23 (table 10-6)
fillers, 211
flash points and ignition temperatures of plastics and packing materials, 218 (table 10-4)
fluorocarbons, 231
fumes and fire gases, 224-25
hazardous properties: of finished products, 215, 217-20; of foam rubber, 223-24; of foamed plastics, 221, 223; in manufacture, 212-14
history of, 201-2
identification of, for fire control, 225-37
manufacture of, 208-10
monomers, 209; combustibility of, 212-13
nylon, 231-32
organic peroxides in manufacture of, 214
phenolics, 232-33
plastic textiles, 229-30
plasticizer, 201, 211
polycarbonate resin, 233
polyesters, 233
polyethylene, 208-9, 210, 233
polymerization, 209
polymers, 209; fire hazard of, 214
polypropylene, 234
polystyrene, 209-10, 234
polyurethane, 234-35
polyvinyl chloride (PVC), 224

Plastics, *(Continued)*
  polyvinyls, 235, 236 (table 10-7)
  pyroxylin plastic, 202
  radiation in manufacture of, 214
  shipping and storage regulations, 237
  silicones, 235-36
  as structural materials, 214-15
  thermoplastic, 210, 211
  thermoset, 210
  toxicity of, 203-4, 207-8, 224-25
  uses of, 211-12
  vinyl plastics, 236 (table 10-7)
  xyloidin, 201
Platinum metals, 196-97
Polycarbonate resin, 233
Polyesters, 233
Polyethylene, 208-9, 210, 233
Polymerization, 209
Polymers, 209, 214
Polypropylene, 234
Polystyrene, 209-10, 234
Polyurethane, 234-35
Polyvinyl chloride (PVC), 224
Polyvinyl, 235, 236 (table 10-7)
Pressurized gas properties, 84-85 (table 5-1)
Pressurized gases, 79-106
  acetylene, 99-103
  anesthetics, 96-99
  behavior of gases, principles of, 79-80
  Boyle's Law, 81, 82
  Charle's Law, 81, 82
  colors of gas containers, 87 (table 5-2)
  Combined Gas Law, 81-83
  cylinder hazards, 89-90
  Dalton's Law, 79n
  emergency measures, 90-91
  flammable ranges of, 89 (table 5-4)
  gas pressure, factors in, 80-81
  Graham's Law, 79n
  hazards of, 87, 89
  Henry's Law, 79n
  hydrogen, 103-4
  identifying type of gas, 80, 81, 83, 86-88
  methane, 92-96
  pressure vs. temperature and volume, 80-83
  properties, 84-85 (table 5-1)
  Roualt's Law, 79n
  reducing pressures, 90
  shipping and storage, 80, 81, 83, 88
Process symbols, chemical, 25
Propane, 33, 109-11
  flame temperatures, 102 (table 5-6)
  flammable range, 89 (table 5-4)
  properties, 34 (table 3-5)
  properties of liquefied, 111 (table 6-1)
  vapor pressures at normal temperatures, 123 (fig. 6-1)
Propionaldehyde, properties of, 58 (table 3-13)
Propyl acetate, properties of, 59 (table 3-13)
Propyl alcohol, 44 (table 3-8)
  properties, 46 (table 3-9)

Propylene glycol
  properties, 47 (table 3-9)
  vapor pressure at normal temperatures, 123 (fig. 6-1)
Protons, 20
Pyrene G-1, 178
Pyrolysis, 4
Pyrophoric, 171
Pyroxylin plastic, 202

R
Radiation, 3
  in plastics manufacture, 214
Radicals, 24, 43-44
  name of, 44 (table 3-8)
Rare earth metals, 197
Red phosphorus, 154
  properties and labeling of, 152 (table 8-2)
  uses of, 154
Refrigerants, 116, 119-20, 126-27
Rhenium, 198
Ring hydrocarbons, 40 (table 3-6), 39-42
Roualt's Law, 79n

S
Salt baths for heat-treating metals, 259-61
  firefighting procedures, 260-61
  overheating, 259-60
  reaction with metals, 260
  reaction with surroundings, 260
  sprinkler protection for, 260-61
  trapped water or air in, 260-61
Saturated hydrocarbons, 33
Service station storage requirements, 74
Silicon, 174 (table 9-1), 192
Silicones, 235
Silver, 198
Smoldering phase, 10-11
Sodium nitrate, 255-59
  firefighting procedures, 259
  flammability, 258-59
  salt baths, 259-61
Sodium peroxide, 254-55
  combustibility, 254
  reactive properties, 254
  toxicity, 255
  uses, 254
Specific gravity, 7
Specific heat, 4-5
Spelter, 193
Spontaneous heating, 149-51
Spontaneous ignition, 9
Steel alloying metals, 191-94
Steel-making, 191
Stibine, 194
Strontium, 181-82
Sulfur, 156-61
  DOT labeling of, 156, 158, 157 (table 8-3)
  dust clouds, 158
  firefighting, 158-59
  flash point, 158, 167 (table 8-4)
  ignition temperatures, 158, 167 (table 8-4)

Tank trucks, 73-74
Tanks and piping, 68-69
Tantalum, 182-84
Tellurium, 193
Temperature. *See also* Critical temperature, liquefied gases; Heat
  definition of, 4
  vs. heat, 4
  and melting and boiling points, 10
  scales, 10
Temperatures at which metals become molten, 173-74 (table 9-1)
Thallium, 174 (table 9-1), 196
Thermal black, 148
Thermite, 190
Thermoplastic, 210, 211
Thermoset, 210
Tin, 174 (table 9-1), 219
Toluene, 41
  fires, extinguishment of, 41
  properties of, 40 (table 3-6)
Toxic fire gases, 12
Toxic metals, 194-96
Triamyl amine, properties of, 56 (table 3-12)
Tributyl amine, properties of, 56 (table 3-12)
Tungsten, 174 (table 9-1), 213
Turpentine, 42
  relative hazards of, 43 (table 3-7)

U
Ultralow-temperature (cryogenic) gases, 148 (table 7-2)
Unsaturated hydrocarbons, 33, 39

V
Valence, chemical, 22
Vanadium, 174 (table 9-1), 193
Vapor cloud, 37, 112
Vapor density, 7-8
Vegetable oils, 61-62
  relative hazards of, 62 (table 3-14)
Vinyl ether
  flammability in anesthetic mixtures, 98 (table 5-5)
  properties and uses of, 52 (table 3-11)
Vinyl plastics, 236 (table 10-7)
Viscosity, 38

W
Water gas, 104
Water
  in carbon disulfide fires, 163
  in cellulose nitrate fires, 207-8
  in coal and charcoal fires, 150-51
  as extinguishing agent, 13-15
  on flammable liquids, 14-15
  on flammable solids, 143, 151
  in hydrogen peroxide fires, 254
  in kerosene fires, 37
  in magnesium fires, 176-78
  on nitrates, 259
  on oxidizing agents, 244
  in phosphorus fires, 152-53, 155
  in plastics fires, 207-8, 237
  reaction on metals, 174
  on sodium peroxide, 254-55
  in sulfur fires, 158
  in tank fires, 66-67
  in titanium fires, 184
  in zinc fires, 194
  in zirconium fires, 187-88
White phosphorus, 151-54
  bombs, 152
  DOT labeling of, 152 (table 8-2)
  fumes, 153
  ignition, 152
  properties of, 152 (table 8-2)
  shipping and storage, 151

X
Xenon, 136
  properties of, 132 (table 7-2)
Xylene, 42
  fires, extinguishment of, 42
  properties of, 40 (table 3-6)
Xyloidin, 201